MULTIVARIATE DATA ANALYSIS
An Introduction

D1189647

The Irwin Series in Marketing

Consulting Editor
Gilbert A. Churchill
University of Wisconsin, Madison

MULTIVARIATE DATA ANALYSIS

An Introduction

BARBARA BUND JACKSON
Graduate School of
Business Administration
Harvard University

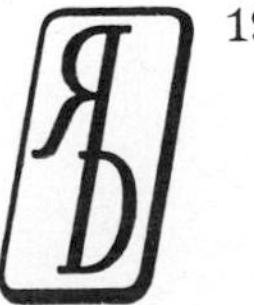

1983

RICHARD D. IRWIN, INC.
Homewood, Illinois 60430

© RICHARD D. IRWIN, INC., 1983

All rights reserved. No part of this publication may be reproduced, stored in a retrieval system, or transmitted, in any form or by any means, electronic, mechanical, photocopying, recording, or otherwise, without the prior written permission of the publisher. The copyright on the following material is held by the President and Fellows of Harvard College, and it is published herein by express permission: AID (Automatic Interaction Detection), Conjoint Measurement, Binary Regression, Discriminant Analysis, Principal Components Analysis, Factor Analysis, Cluster Analysis, Multidimensional Scaling, Likelihood, The Analysis of Variance. Copyright © 1980 by the President and Fellows of Harvard College. This material was prepared by Professor Barbara B. Jackson as a basis for class discussion.

ISBN 0-256-02848-6

Library of Congress Catalog Card No. 82–83832

Printed in the United States of America

1 2 3 4 5 6 7 8 9 0 MP 0 9 8 7 6 5 4 3

For Henry Bund 1912–1966

Preface

This is a book about techniques for data analysis—methods that can help investigators extract the information contained in a set of data. Its aim is to give users information they need for the sensible application of these techniques in the exploration of databases. There are two main audiences for whom the material is most appropriate. One consists of people not already familiar with the techniques, who could use the methods profitably in their work (as managers, students, or researchers). For this group, the following chapters provide an introduction to a set of important multivariate methods. The second group consists of people who have studied or worked with some of these methods. For this group, the chapters will provide a refresher with a different emphasis than most treatments—one that stresses practical use of the procedures, as well as their limitations.

Throughout, the book takes the point of view of users (rather than the view of the more technical specialists who provide the computer programs to implement the procedures). The book's primary goal is to help the reader learn as much as possible from sets of data, using rigorous analysis with well-established methodologies. Its underlying philosophy is that good data analysis involves looking at a database from different angles and with different tools, using all appropriate data analytic techniques to leverage the judgments and insights of an investigator. The aim of such analysis is to extract as much insight as possible from a rather precious resource—a good, useful set of data. Such analysis provides the best foundation for whatever comes next—in some cases for further data collection, in some for further pursuit of a research topic in other ways, and in some for sensible, effective managerial decisions and actions.

The investigator's purpose in analyzing data is central in determining what methods should be used to make sense of any particular set of data. For example, the extremely rich database of U.S. census information is used by many different investigators to pursue a wide variety of questions in demography, marketing, and other areas. It is particularly important for users of multivariate analysis to understand what various techniques do and don't accomplish, when they're useful, and when they're not. Hence, this book stresses the objectives of the multivariate procedures.

After the initial background chapter, each chapter covers one general technique. Each focuses on the *basic idea* of a technique—when and why it is useful in the analysis of data. Each covers the *inputs* that the user must provide—in the form of input data and also in the form of choices of criteria, selection of methods, or other decisions the user must make. Each chapter also emphasizes the interpretation of results or *outputs*—what can, as well as what cannot, be learned from the procedures.

The book assumes that computers will perform the actual computations required to apply the techniques. Thus, users need not know all of the computational details—and they certainly do not need to know the details of computational shortcuts that make computer programs for multivariate analysis more efficient. Consequently, the chapters include computational detail only insofar as it is useful for either of two purposes.

First, computations are sometimes included to illustrate definitions. Many people find it easier to understand the concept of a residual, or of R^2, or of a specific method for cluster analysis if they follow through the details of a specific (small-scale) example. Where practical, the chapters include such illustrative computations.

Second, there are instances in which users of multivariate techniques must know about some of the computational details in order either to specify inputs or to interpret outputs. For example, they must know a bit about the computations used in nonmetric multidimensional scaling in order to understand what it means to specify a starting configuration for the procedure. Hence the chapters must include some details about computation for this second reason as well.

The chapters stress examples, illustrating applications of the various methods. Most of the examples relate to marketing (since that is the field in which I work), but the techniques are used and useful in a wide range of fields and the book is appropriate for users in many areas beyond marketing. Readers should be able to use the examples in the chapters as guides in identifying interesting examples from their own fields for application of the multivariate procedures.

The discussions assume that readers are familiar with the basic ideas of cross-tabulation and regression (which are covered in a wide variety of books on data analysis and statistics). Chapter 1 summarizes the background concepts needed for the other chapters, but it is neither fully detailed nor complete in its scope. Chapter 1 thus will not be sufficient for readers who do not have other background in cross-tabs and regression.

The remaining chapters cover a selection of the most widely used multivariate techniques (with the exception of procedures for time series analysis, which are covered in numerous books on econometrics and statistics). They can be read in any order. Appendix A presents material necessary for the chapters on binary regression and on discriminant analysis. Appendix B presents a brief comparison of the techniques, which some readers will find a useful addition to the individual chapters.

Throughout, the book strongly emphasizes data analysis. It is intended to help users extract the information in their databases in exploratory phases of analysis. It does not consider formal inferential procedures for using a sample to draw formal inferences about a wider population or process.[1]

The reason for this strong orientation is that I believe it is a correct one for much of the work in marketing and other applied fields. Learning as much as possible from a set of data is both challenging and important. The process facilitates both good research and effective managerial action. To be sure, there are cases in which it is very sensible for investigators to try to determine whether conclusions tentatively reached with one set of data can be applied to additional fresh data. In such cases, the rigorous procedures typical of formal inference are quite appropriate.

The requirements for formal inference are considerable, however, especially with regard to data. For example, in correct classical testing procedures, investigators may not use the same data to test multiple hypotheses without very substantial tightening of the levels they use as hurdles in testing. The basic procedure is one of stating an hypothesis, collecting data to test it, running one test, and then throwing the data away.[2] The process of data analysis, by contrast, involves mining a database to try to uncover as many insights as possible. In such investigations it is good, not bad, to look at a database from different angles and with different tools, to try to learn as much as possible from it. The flavor in such exploratory work is that of extracting as much insight as one can from the resource—the data.

This type of analysis can be viewed as the use of data analytic techniques to leverage the judgments and insights of an investigator—a researcher and/or a decision maker. Good, useful results are produced

---

[1]The chapters on binary regression and discriminant analysis discuss prediction for observations outside the original database because such prediction is so natural a part of the use of those methods—but the emphasis is still on exploratory, data analytic work, not on formal inference.

[2]A related argument can be made about formal inference by Bayesian investigators.

when investigators combine good insight and judgment with sound application of the methods. In good data analysis, this combination produces considerably more learning than could either judgment or technique alone. Thus, the chapters that follow note the ways in which judgment enters analysis with the multivariate procedures.

Because the use of judgment is not applauded or fashionable in the academic journals in marketing and related fields, the assumption of this book (that judgments and formal procedures do—and should—interact) deserves a bit more attention. Many investigators in such fields work hard to ensure that the analyses they perform on data would be replicated exactly by other (independent) investigators working with the same techniques on the same sets of data. Those investigators apparently believe that such procedures bring objectivity to their work; they feel the more-judgment-based analyses are less objective, less scientific, and less desirable.

At least some judgment is required in any data analysis (in the construction of a database, for example, or in the choice of technique). In addition, mechanically (or "objectively") defined methods can usually not be defined to do exactly what we want done in data analysis (to the extent that we know at the outset what it is that we want done). Further, good investigators have good insights and judgments. I believe those insights and judgments should be exploited rather than avoided.

Thus, quantitative analysis is better considered a way of aiding or leveraging judgment rather than as a way of replacing it. The best way to learn what a set of data can teach is to combine judgment and technique, in a manner tailored to the database and to the purpose of the analysis, to form a useful and rigorous whole.

This emphasis on combining judgment and technique is also the key to getting managers to accept and apply results. Managers are often suspicious of results of analysis that do not incorporate insights and judgments. At the same time, they are increasingly aware of (and, in some cases, familiar with) newer and more formal tools. Combining judgments and technical procedures to give rigorously reasoned analyses is an effective way to apply both—and to get them used.

While they will need to consult the following chapters for more complete discussions of the techniques, readers may find these brief descriptions a useful start:

Chapter 2: AID—an automated procedure for cross-tabs analysis (essentially, a "fishing expedition").

Chapter 3: Conjoint measurement—a procedure much like additive cross-tabulation, but with an ordinal (rather than cardinal) dependent variable. Most commonly used to explain preference rankings.

Chapter 4: Binary regression—techniques for regression with a binary (dummy) dependent variable.

Chapter 5: Discriminant analysis—a technique for calibrating a procedure to classify individual observations into one of a set of groups.

Chapter 6: Principal components analysis—a procedure for restating the information in a particular set of observations on one set of variables in terms of an alternate set of variables.

Chapter 7: Factor analysis—a technique for analyzing the internal structure of a set of variables—for describing the linkages among a set of observed variables in terms of unobservable, underlying constructs called factors.

Chapter 8: Cluster analysis—a collection of procedures for grouping entities (observations or variables) that are similar to one another.

Chapter 9: Multidimensional scaling—procedures for converting input on a single measure of dissimilarity (or similarity) among objects into geometric representations of those objects in multiple dimensions.

Acknowledgments: These chapters were developed when I taught a doctoral seminar on multivariate methods at Harvard Business School in 1978 and 1980. I would like to thank the students in the 1978 course (who made do without proper written notes) and the students in the 1980 course (who helped with improvements to the first version of the materials). I would also like to thank Professors Paul Vatter and Arthur Schleifer for their careful reading and comments on the materials and Professors Robert Glauber, Arthur Schleifer, and, especially, Robert Schlaifer for the computer programs with which I was able to explore the multivariate procedures. Professor Schlaifer's documentation of the AQD collection of programs for data analysis was an especially important resource; moreover, it was his initial work and careful thought about multivariate methods that interested me and others at Harvard in the area and suggested topics of particular interest. Finally, I am very grateful to Frances Charon for her exceptional work in typing the manuscript.

Barbara Bund Jackson

Contents

CHAPTER 1

Basic Concepts

The following chapters assume that the reader understands the basic tools of multivariate analysis: concepts of variables and observations, distributions, data analytic models, cross-tabulations, and ordinary regression. This introductory chapter reviews those ideas briefly. The review will not be sufficient for readers who have never been exposed to the basic ideas. Further, it does not cover the basic concepts in depth but merely summarizes ideas and points that will be especially useful for the chapters that follow. The review is important, however, both because it covers fundamental ideas and because this chapter introduces some notation and terminology which will be used in later chapters. Therefore, readers will want to read through this chapter before proceeding to any of the following ones; in addition, they will likely want to return to this introductory material for reference as they consider the following topics.

DATABASES, OBSERVATIONS, AND VARIABLES

We assume that the investigator planning to use one or more of the multivariate techniques has a set of data or a *database* consisting of information about a group of *observations*. The information consists of values of a set of *variables*. For example, we might have a database with 50 observations, one for each state in the United States. Variables might be population, geographical area, and so on. Or we might have a database in which there was one observation for each of a sample of consumers and there were variables for that consumer's weekly expenditures in each of five categories of food products: fruit and vegetables, poultry and meats, dairy products, bread and grains, and other. In general, it is useful to think of observations as entities—people or things; variables are characteristics of those entities.

The very strong orientation of this book is toward data analysis (the study of a particular set of data) rather than toward statistical inference (the use of formal methods to use information from a sample to draw formal inferences about the population or process from which that sample came). In inference, of course, investigators must worry about issues of *bias*—of samples that are not representative of the underlying populations or processes. For example, formal inference on a sample from a telephone survey before the U.S. presidential election in 1936 predicted that Roosevelt would lose the election; the analysts neglected the fact that telephone subscribers were by no means a representative sample of the U.S. electorate in 1936. Similarly in inference investigators must also worry about *errors* in their data.

Despite their emphasis on specific samples, data analysts must also worry about bias. Often, although not always, they will want to use their samples to suggest more general hypotheses even though they do not perform formal statistical inference. In addition, they must worry about such issues as errors in measurement of their variables or errors in transcription of values.

Thus, data analysts should generally begin with a careful consideration of their data to try to identify sources of bias and errors. If they find that their databases are seriously biased, they may end their analyses without proceeding further. If a database is not so biased as to be useless, it may still contain some observations that are not useful—because of errors in recording the data perhaps, or for some other reason. To identify at least the most obvious of the bad observations, analysts examine their databases for *extreme values*, or *outliers*. An extreme value of a particular variable is a value of that variable, on some observation, that is much larger or much smaller than the values taken by that variable on other observations. Extreme values can occur because of errors in measurement or transcription and also because inappropriate observations were included in the database (for example, an adult might have been included by mistake in a set of data about elementary school children). On the other hand, extreme values can occur in perfectly good data simply because the values of some variables are in fact very large or very small on one or a few observations. For example, a surprisingly high income figure may be an error but it may also simply be the correct value for someone who quite properly belongs in the group under study but who has a surprisingly high income. The analyst's job is to identify extreme values and then to try to determine whether those values are errors or not. If they represent errors, analysts will generally remove the affected observations from the databases. If, on the other hand, an extreme value is not an error, the analyst would be throwing out useful (and often important) information by discarding it; such observations should be kept.

Another aspect of database structure that runs through the following chapters is the issue of size—the number of observations in the database. In general, analysts will feel (and be) safer in basing their analyses on larger rather than smaller numbers of observations (provided, of course, that those observations represent useful data). To be sure, there are cases in which some entire group of observations of interest is very small and then useful analysis would consider a small set of data, but in general, analyses based on few observations are dangerous. Results tend to be highly idiosyncratic to the particular observations included. The question of how many observations are enough depends on the particular techniques used and on the purposes of the investigator.

TYPES OF VARIABLES

Cardinal variables take values which we can meaningfully manipulate through addition or subtraction, which we can compare quantitatively, and so on. *Ratio variables* can meaningfully be multiplied or divided as well. For example, hourly pay would be such a variable. A worker earning $10 per hour earns exactly twice the pay of one earning $5 per hour; the pay differential between rates of $12 and $14 is exactly the same in dollars as the difference between $9 and $11; and so on. *Interval variables*, such as temperature, can be added or subtracted but not multiplied or divided (because the zero values for such variables are chosen arbitrarily). In this book the term "cardinal variables" generally means either ratio or interval variables.

Ordinal variables contain information about rank or order but they do not possess full cardinal properties. An example is a scale running from 1 (most favorable) to 10 (least favorable) on which respondents might rate possible new print—ads for a product. The variable has an ordering—1 is better than 2 or any higher number, 2 is better than 3, and so on. Yet, it is not at all necessarily true that the difference in ratings for some individual between a 2 and a 4 is the same as the difference for that individual between a 5 and a 7. Similarly, we cannot say that a 4 rating is half an 8 rating, and so on.

Categorical variables take values corresponding to categories but do not possess either cardinal or ordinal properties. For example, a regional variable might classify geographical locations into far west, southwest, southeast, north central, and northeast. Or, a variable for occupation might classify workers as professional, clerical, skilled manual, and unskilled manual. *Dummy variables* can be considered a subclass of categorical variables; dummy variables take only two possible values.

In representing variables for computer (or other) analysis, analysts often *recode* (or restate) variables. For example, most computer programs require that variables have numbers as values—even though those numbers may simply be shorthand for longer category names. In such cases, we assign numbers to the categories. We might, for example, use 1 for far west, 2 for southwest, 3 for southeast, 4 for north central, and 5 for northeast. Note, however, that the underlying variable remains a categorical one; it does not make sense to perform arithmetic operations or comparisons to determine relative size on the coded values of the variable. Because the codes used for this purpose do not have cardinal or even ordinal usefulness, the precise nature of the coding can easily be varied. For example, dummy variables are most commonly coded 0 or 1. This choice is especially helpful when the variable says whether or not an observation has some property or condition (is or is not a home-owner, for example) and the 0 is used to code the absence of the condition, with 1 coding its presence. On the other hand, it is also perfectly possible to code a dummy variable with the values 1 and 2 instead of 0 and 1.[1]

Sometimes analysts find it useful or necessary to recode the values of variables that already have quantitative values. For example, investigators might be using a variable for annual income. The original variable is fully cardinal. The investigators might want to recode the variable to represent income ranges and might use 1 to code values less than $15,000, use 2 for values at least $15,000 but no more than $25,000, use 3 for values at least $25,000 but no more than $40,000, and use 4 for values $40,000 or higher. The new (recoded) variable is ordinal but not cardinal. It is also *few-valued* rather than *many-valued.* Some multivariate techniques require that some or all of the variables used must be few-valued. (For example, in cross-tabulation, as described below, independent variables must be few-valued.)[2]

SUMMARIES OF THE VALUES OF A VARIABLE

Analysts use a variety of tools to summarize the information in the values of a particular variable. For example, suppose that we are work-

---

[1]Some computer programs require that variables be coded in specific ways—starting at 0 perhaps or starting at 1. Such requirements are peculiar to the programs, however, and not inherent in the idea of coding. While users must certainly learn and follow the conventions of the particular programs they are using, the conventions should be viewed as properties of the programs.

[2]The question of how many values a variable can take and still be sufficiently "few-valued" depends on the number of observations in the database, on the particular techniques used, and on the purposes of the investigator.

TABLE 1–1

Grade	Count (Number of Students)
54	1
57	0
60	3
63	4
66	2
69	7
72	18
75	20
78	16
81	9
84	11
87	3
90	4
93	1
96	1

ing with a database containing 100 observations, each corresponding to a student in a specfic course. Suppose that the grading system used in the course allows grades ranging from 54 to 96, inclusive, but using only every third number. (Thus, the possible grades are 54, 57, 60, and so on.) Suppose that the database contains a variable for grade. With 100 observations there will be 100 values for the grade variable. An initial simple way to summarize the information in the variable is to count the number of observations for each possible grade value and then to produce a table of grades and *counts*. Table 1–1 presents such information. This type of presentation will be a more succinct statement of the information contained in a variable which takes the same value, for at least a few of its possible values, on multiple observations.

Next, we can produce a *frequency distribution* which gives for each value of the variable the fraction (rather than the count) of observations on which the variable takes that value. Since the data summarized in Table 1–1 involve 100 observations, the counts are divided by 100 to give frequencies. The results are shown in the third column in Table 1–2. The last column in that table gives *cumulative frequencies;* for each possible grade value it gives the fraction of observations in the sample on which the variable's value was no larger than the specified value.

Thus, frequencies and cumulative frequencies summarize values taken by a particular variable in a sample in much the way that a *probability density function* and a *cumulative distribution function* summarize information about a theoretical or a population distribution of val-

TABLE 1–2

Grade	Count	Frequency	Cumulative Frequency
54	1	.01	.01
57	0	.00	.01
60	3	.03	.04
63	4	.04	.08
66	2	.02	.10
69	7	.07	.17
72	18	.18	.35
75	20	.20	.55
78	16	.16	.71
81	9	.09	.80
84	11	.11	.91
87	3	.03	.94
90	4	.04	.98
93	1	.01	.99
96	1	.01	1.00

ues. For example, Figure 1–1 shows a probability density function; in such a figure, the height of the function above a range of values is proportional to the probability of a value from that range. Figure 1–2 gives the corresponding cumulative distribution function; for any specific value on the horizontal axis, the function gives the probability of a value no greater than that specific value. (It is worth noting that the use of the term *frequency* rather than *probability* for sample values is not universal. *Probability* is sometimes used for sample fractions.)

Fractiles are determined from cumulative frequencies or cumulative probabilities. If f is a fraction between 0 and 1, then the f fractile divides a distribution as nearly as possible into two parts containing f and $1 - f$ of the frequency or probability. In Table 1–2, the grade value 75 is the .5 fractile or *median*. Similarly, 72 is the .25 fractile and 84 is the .9 fractile. The easiest way to find fractiles is often to read values from graphs of cumulative frequencies or probabilities. For example,

FIGURE 1–1

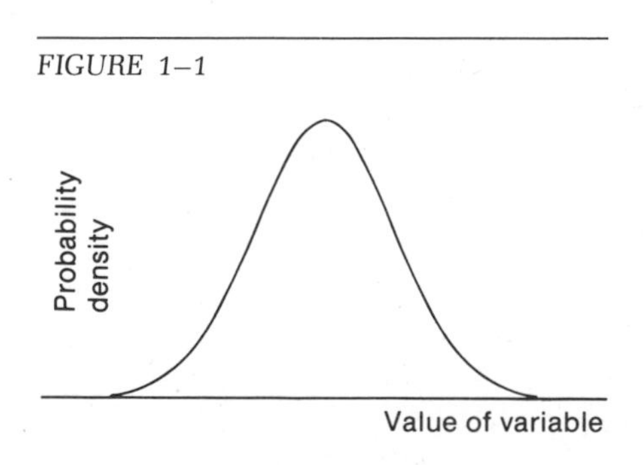

FIGURE 1–2

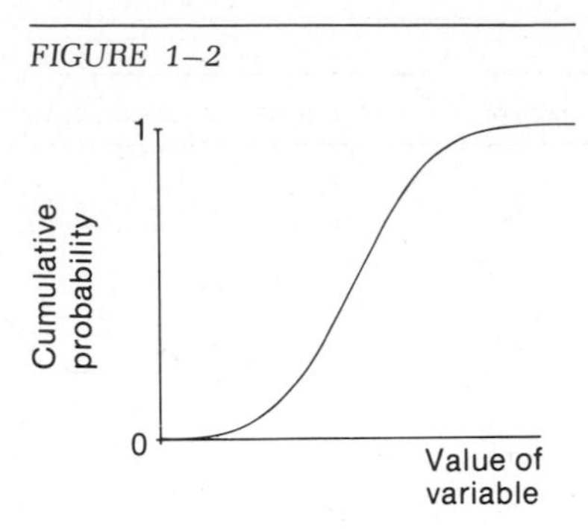

Figure 1–3 shows how to read the .3 fractile from the graph in Figure 1–2. One finds .3 on the vertical axis, reads across to the curve, and then straight down. The intersection with the horizontal axis gives the desired fractile (labeled x* in Figure 1–3).[3]

The *mean, variance,* and *standard deviation* are other summaries of the values of a single variable. Suppose we call the variable y. Then the mean of y, denoted $M(y)$ in this book, is the average of the values of y. Table 1–3 presents values of the Dow Jones Industrial Average at the end of each of 10 weeks. It also shows the calculation of the mean $M(y)$.

The variance of y, which we will denote by $V(y)$, is the average value of the squared difference between y and $M(y)$. In symbols,

$$V(y) = M\{[y - M(y)]^2\}$$

Table 1–4 demonstrates the calculation of the variance of the values from Table 1–3. The standard deviation of y is defined as the square root of $V(y)$—in Table 1–4, 22.6.

FIGURE 1–3

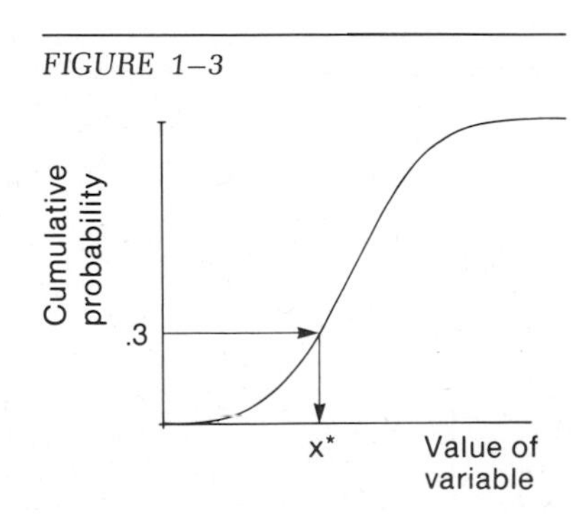

[3]In a graph of cumulative frequencies, analysts may read across to find a flat portion in the graph. In such cases, it is conventional to define the fractile as the horizontal value at the midpoint of the flat region.

TABLE 1–3

Observation Number	Dow Jones Industrial Average
1	922
2	954
3	928
4	895
5	931
6	894
7	920
8	889
9	880
10	892
	Sum = 9,105
	$M(y) = 9{,}105/10 = 910.5$

In some multivariate analyses, investigators use *standardized variables,* defined on the basis of their original variables. To standardize a variable y, on each observation we subtract $M(y)$ from the value of y and then divide that difference by the standard deviation of y. For example, in Table 1–4 the standardized value for the first observation is $(922 - 910.5)/22.6$, or .51. Standardized variables always have mean 0 and variance 1.

MARGINAL AND CONDITIONAL DISTRIBUTIONS

When dealing with more than one variable, we can consider both marginal frequencies (or distributions) and conditional ones. The *marginal*

TABLE 1–4

Observation Number	y	$y - M(y)$	$[y - M(y)]^2$
1	922	11.5	132.25
2	954	43.5	1,892.25
3	928	17.5	306.25
4	895	−15.5	240.25
5	931	20.5	420.25
6	894	−16.5	272.25
7	920	9.5	90.25
8	889	−21.5	462.25
9	880	−30.5	930.25
10	892	−18.5	342.25
			Sum = 5,088.5
		$V(y) = 5{,}088.5/10 =$	508.85

TABLE 1–5

Soft Drink	Count	Frequency
A	45	.265
B	50	.294
C	75	.441

distribution of a variable gives the frequencies or probabilities of different values of a variable, considered alone. For example, we might have data on the choices made by 170 people when asked to select their favorite from among soft drinks A, B, and C. Table 1–5 presents marginal frequencies for preference.

Suppose, however, that we have additional information about each of the 170 respondents—in particular, that we have each person's age range, with possible ranges 20–29, 30–39, and 40–49. We might try to determine whether the distribution of preferences was the same for each of the three age groups. For that purpose, we would use *conditional distributions* of preference given age. In other words, we would look at frequencies of choice within each age group. Table 1–6 presents the conditional distributions. Notice that the frequencies in each column total 1.0; each column gives a complete distribution, conditional on a particular age range.

Two variables are called *statistically independent* if the marginal distribution of one of them is identical to the conditional distribution of that same variable given (or conditional on) each possible value of the second variable. The two variables in Table 1–6 are not statistically independent; they would be called *statistically dependent*. In contrast, the counts and frequencies given in Table 1–7 describe statistical independence in similar preference data.

If two variables are statistically independent, then knowing the value of one of them tells us nothing new about the likely values of the other. In Table 1–7, a respondent is just as likely to choose fruit juice G in either age group; knowing a respondent's age does not change our expectations about his or her choice of drink. In Table 1–6, on the other

TABLE 1–6

	Counts by Age			Conditional Frequencies		
Soft Drink	20–29	30–39	40–49	20–29	30–39	40–49
A	15	10	20	.250	.200	.333
B	20	15	15	.333	.300	.250
C	25	25	25	.417	.500	.417
				1.000	1.000	1.000

TABLE 1–7

Fruit Juice	Counts, by Age		Conditional Frequencies	
	10–29	30–49	10–29	30–49
E	25	20	.25	.25
F	25	20	.25	.25
G	50	40	.50	.50

hand, the marginal chance that a respondent would choose A is .265. If we know the respondent is 40–49, however, the chance that s/he chooses A is .333. Thus, with statistically dependent variables, knowing the value of one does give us more information about likely values of the second.

Different notation is used to distinguish marginal and conditional frequencies (or probabilities) from one another. For example, we might use $pr(A)$, $pr(B)$, and $pr(C)$ (or $p(A)$, $p(B)$, and $p(C)$) to denote the marginal probabilities of choices of A, B, and C. To show conditionality, we generally use a vertical line within the probability symbol. For example, $pr(A|20\text{–}29)$ (or perhaps, $p(A|20\text{–}29)$) would be read as the probability of A, given that (or conditional on the fact that) the respondent was in the 20–29 range of age.

Correlation and *covariance* are related measures of a particular form of statistical dependence; they measure the extent of linear relation between two variables. If y and x are the two variables of interest, then the covariance between y and x is defined as the average value of the product of the difference between y and $M(y)$ and the difference between x and $M(x)$. In symbols, the covariance is $M\{[y-M(y)] * [x - M(x)]\}$. The correlation is the covariance divided by the product of the standard deviations of the two variables; it is also the covariance between standardized versions of the variables. Correlations always lie between -1 and 1. Variables with 0 correlations are said to be *uncorrelated*. Correlated variables are also sometimes called *collinear*. A group of several correlated variables is said to show *multicollinearity*. Correlation is not the only form of statistical dependence; two variables can be uncorrelated but still not statistically independent.

DATA ANALYTIC MODELS

A number of the techniques for analyzing multivariate data involve a *dependent variable*, one variable that is singled out to be explained by other variables. The variables that do the explaining are most commonly called *independent variables*, although *explanatory variables* is a more descriptive and less confusing term; both terms are used in this book. (The confusion about the term *independent variables* arises in large part because of the separate concept of statistical independence. Two independent variables—in a cross-tabulation or regression, for example—need not be statistically independent, although they could be. The concepts of an independent variable and of statistical independence are totally different ones.)

Many data analytic *models* decompose observations into a *predicted value* and a *residual:*

$$y = \text{Predicted value} + e$$

where y is the value of the dependent variable and e is the residual. Often, we decompose observations already in the database, predicting them only retrospectively, and hence the term *explained value* might be better than *predicted value* but predicted value is common. Sometimes $\hat{y}$ (read, "y hat") is used for the predicted value of a model.

The simplest such model is

$$y = M(y) + e$$

where $M(y)$, the mean of the dependent variable, is used as the predicted value (and there is no independent variable involved). This simplest model gives a standard for comparing results of explaining values of the same y with more complex predicted values. For such comparisons, we need to use a *criterion*, or *measure of fit*, of the model to a set of data. The most common such criterion is the average squared error (or average of the squared residuals) when the model is applied to the database. This *squared-error criterion* weights large errors particularly heavily (since they are squared) and, thus, it is particularly sensitive to outliers, or extreme values, in the data. (Other possible criteria, such as the average of the absolute values of the residuals, are less sensitive to extreme values, although they are far less frequently used.)

Table 1–8 demonstrates the decomposition of a variable into mean plus residuals according to this simplest model. The total squared error when divided by the number of observations gives the average squared error, which is also the variance of the residuals, or $V(e)$. (This value is a variance because the average residual with the mean used as the predicted value is 0.) For this model, $V(e)$ is equal to $V(y)$, the overall variance of y. Thus, $V(y)$ is the standard of comparison for measures of fit of other models, if we are using a squared-error criterion. Suppose that we have used some other model to find predicted values and residuals (perhaps a cross-tabulation model or a regression model) and that we have found the corresponding residuals and $V(e)$ for this new

TABLE 1–8

Observation Number	y	Predicted Value $M(y)$	Residual e	e^2
1	164	173.75	− 9.75	95.06
2	172	173.75	− 1.75	3.06
3	178	173.75	4.25	18.06
4	175	173.75	1.25	1.56
5	193	173.75	19.25	370.56
6	166	173.75	− 7.75	60.06
7	155	173.75	−18.75	351.56
8	191	173.75	17.25	297.56
9	197	173.75	23.25	540.56
10	168	173.75	− 5.75	33.06
11	171	173.75	− 2.75	7.56
12	149	173.75	−24.75	612.56
13	164	173.75	− 9.75	95.06
14	177	173.75	3.25	10.56
15	182	173.75	8.25	68.06
16	185	173.75	11.25	126.56
17	156	173.75	−17.75	315.06
18	163	173.75	−10.75	115.56
19	170	173.75	− 3.75	14.06
20	168	173.75	− 5.75	33.06
21	178	173.75	4.25	18.06
22	176	173.75	2.25	5.06
23	177	173.75	3.25	10.56
24	195	173.75	21.25	451.56
			Sum =	3,654
			V(e) =	3,654/24
			=	152.25

model. Then, a common measure of the fit of this new model is R^2, defined by

$$R^2 = \frac{V(y) - V(e)}{V(y)}$$

In this equation $V(e)$ is the average squared residual for the new model and $V(y)$ is the variance of y, which is also the average squared residual for the model using only the mean for prediction. R^2 values always lie between 0 and 1. Higher values are better. An R^2 of 1 indicates that the model has explained all of the variability in y, reducing the residuals to 0. An R^2 of 0 says that the model has explained none of the variability in y.

Cross-tabulation models, regression models, and other explanatory models can be viewed as ways of improving on the predicted value portion in the basic model equation above. In general, those models define the predicted value on the basis of values of independent variables, using the database as a guide for the precise values for the prediction equation. Analysts must always bear in mind a few warnings about such models. In particular, the models measure *association* in the database; they do not measure *causation*. Most statisticians and data analysts have their own favorite examples of association that does not seem to be causation—for example, the miles of interstate highway in Texas and the divorce rate in that state have increased together over time. We would therefore get good measures of fit for a model that predicted divorce rate on the basis of miles of interstate highway. Such a model, however, has several problems. For one thing, it does not help us understand the factors causing high divorce rates—and if the explanatory variable were a bit more plausibly connected with the dependent one we might be tempted to hypothesize such a causal relationship when there was only a statistical association. In addition, statistical associations do not necessarily persist over time. If the government cut off funds for the highway system so the miles in Texas did not increase, it would be unlikely that the divorce rate would behave similarly. If the relationship between an independent variable and a dependent one is causal, we would in general (although not always) expect that relationship to persist more reliably.

Hence, we would often be far more comfortable (as well as better informed) if our models contained causal relationships. Models cannot prove causality, however. It is the job of the analyst to explain why a particular association, demonstrated by some model, might also be a causal relationship.

A related concept is the idea of a *proxy variable*. Suppose that we know, somehow, that variable x is an important causal influence on variable y. Suppose that we are trying to explain y. Suppose, in addition, that x is associated (though not in a causal way) with a third variable z. If so, then a model for predicting y from z will give good results, even though z does not cause y; z is serving as a proxy, or stand-in, for x. While proxies are often useful in data analysis (especially when we cannot obtain measures of what we feel is the true causal variable), their use involves the dangers inherent in modeling based on association rather than causation—and, again, the analyst rather than the computer is responsible for making judgments about the form of relationship.

The degree of danger in building models based on association or proxy variables rather than on causation varies, depending on the purpose of the analysis. Sometimes analysts are most concerned with overall results, or predicted values—when, for example, they are trying to prepare *forecasts* or predictions with a model. In such cases, unless the form of an association or the relationship of a proxy changes, models involving proxy variables are often very useful. On the other hand, analysts are sometimes concerned with understanding *structural relationships*—the actual relations between some independent variables and a dependent variable. In such cases, analysts would be more concerned with using what they believed were true causal variables.

In these cases, the analysts must worry especially about the possible existence of what are called *intervening variables*—variables that are related to both the independent and the dependent variable and that get in the way of allowing the analysts to see clearly the relationship of interest. As a simplified but clear example, suppose that we are planning to introduce a new, slightly higher-priced item at one of two fast-food restaurants in a chain. We want to choose the location, A or B, that would be most favorable for a test. Suppose that for some reason we have data only on the total amount of each tab in the two locations, and suppose that we find that the average tab at A is higher than the average tab at B. Can we conclude that patrons at A are apt to be more receptive to higher-priced items? Not, for example, if location A is one that is more likely to attract larger groups—if, for example, it is located near an amusement park that is frequented by families with children, while B attracts smaller parties. We may find that size of party is an intervening variable that is related both to location and to size of tab. In fact, patrons at location B might spend more per person but their total checks might be smaller because the average party size at B was smaller. Thus, location B might be the better site for the test. In this example, the analyst would be far safer considering the average cost of a single meal at A and at B, thus eliminating the confusion caused by differences in size of party.

This example is a clear one (and the reader will likely feel that most sensible people would not make the mistake of using total check amount uncorrected for number of people). In other cases, however, the issue of possible intervening variables is a serious one, and the job of identifying such variables falls to the analyst, not the computer. Good analysts will spend considerable thought on the issue of what variables could be masking or distorting the relationships they care about. The issue requires considerable judgment.

Some data analytic techniques, including so-called *stepwise procedures,* include computer searches for that set of independent variables in some general type of model, such as a regression model, that gives the best fit to the data at hand. Such procedures are often called *fishing expeditions* (because they fish for results). Such procedures are always dangerous and should be used only with extreme care, if at all. Just as most people can find all kinds of meaningful patterns in what are really random data if they look long enough, so the computer is highly likely to uncover results on a fishing expedition that reflect idiosyncracies of the particular data at hand rather than a meaningful model for sensibly explaining the dependent variable. The attraction of such techniques is likely that they require little judgment from the investigator—but that, of course, is also their major shortcoming. Good data analysis generally requires good judgment.

CROSS-TABS

Cross-tabulation or *cross-tabs* models are powerful simple tools for analysis of data. In the simplest cross-tabs model, values of one independent variable are used to define *cells;* there is one cell for each *level* of the independent variable. (Sometimes, a potential explanatory variable is many-valued rather than few-valued. In such cases, the many values are summarized into a few levels, often by dividing values of the original variable into ranges.) Observations are then sorted into the cells on the basis of their values on the independent variable. The predicted values for this model are the *cell means*—means of the values of the dependent variable in each cell. We denote the mean of the values in the ith cell by $M_i(y)$. The equation for this simple cross-tabs model is

$$y = M_i(y) + e$$

Thus, the predicted value for an observation is the mean of the cell into which that observation is sorted on the basis of the independent variable.

Table 1–9 contains some hypothetical data on monthly heating costs for four different types of construction of single-family homes in

TABLE 1–9

Observation Number	Monthly Cost ($)	Construction Type
1	164	1
2	172	1
3	178	2
4	175	3
5	193	3
6	166	4
7	155	4
8	191	2
9	197	2
10	168	1
11	171	3
12	149	4
13	164	4
14	177	1
15	182	2
16	185	2
17	156	1
18	163	3
19	170	4
20	168	4
21	178	3
22	176	3
23	177	2
24	195	1

TABLE 1–10

Type 1		Type 2		Type 3		Type 4	
164	177	178	182	175	163	166	164
172	156	191	185	193	178	155	170
168	195	197	177	171	176	149	168
$M_1(y) = 172$		$M_2(y) = 185$		$M_3(y) = 176$		$M_4(y) = 162$	

the same area. The dependent variable will be monthly cost and the independent variable will be construction type. Table 1–10 shows the same data sorted into four cells; it also shows the four cell means. Table 1–11 shows the results of applying the cross-tabs model to these data. Each observation is "predicted" by its cell mean. $V(e)$ is the averaged squared value of the resulting residuals, or 84.08. For this set of data $V(y)$, the variance of the original values of the dependent variable, is 152.25, as shown in Table 1–8, which used the same y. Therefore, R^2

TABLE 1–11

Observation Number	y	Predicted Value	e	e^2
1	164	172	− 8	64
2	172	172	0	0
10	168	172	− 4	16
14	177	172	+ 5	25
17	156	172	−16	256
24	195	172	+23	529
3	178	185	− 7	49
8	191	185	+ 6	36
9	197	185	+12	144
15	182	185	− 3	9
16	185	185	0	0
23	177	185	− 8	64
4	175	176	− 1	1
5	193	176	+17	289
11	171	176	− 5	25
18	163	176	−13	169
21	178	176	+ 2	4
22	176	176	0	0
6	166	162	+ 4	16
7	155	162	− 7	49
12	149	162	−13	169
13	164	162	+ 2	4
19	170	162	+ 8	64
20	168	162	+ 6	36
				Sum = 2,018
				V(e) = 2,018/24 = 84.08

for this model is

$$R^2 = \frac{152.25 - 84.08}{152.25} = .45$$

Such a cross-tabs model divides the original variance $V(y)$ into two parts. The first, or explained, portion is the variance of the values in the predicted value column in Table 1–11; it is denoted $V[M_i(y)]$. The second part is the variance of the residuals from the model. Thus

$$V(y) = V[M_i(y)] + V(e)$$

Cross-tabs models can be defined with more than one independent variable. With two independent variables, we use values of both independent variables to sort observations into cells. We might, for example, use weekly expenditures on groceries as the dependent variable, with family size (number of people) and family income (coded into five ranges) as the independent variables. Suppose that our data considered families of 2, 3, 4, or 5 people only. Then we would consider the cells shown in Figure 1–4. We would sort each of our observations into one of those cells and would then calculate the mean expenditure for the observations in each cell. If we denote the cell mean for the ith level of people and the jth level of income by $M_{ij}(y)$, then the model would be

$$y = M_{ij}(y) + e$$

We would calculate R^2 for this model much as we did in Table 1–11 for the previous model.

We may find in a cross-tabs model that the prediction is just about as good in one cell as in another—in other words, that the variability of the values in one cell around their cell mean is just about the same as the variability of the values in another cell about their cell mean. In such situations we say that our data display *homoscedasticity*. If, on the other hand, the variability of values of the dependent

FIGURE 1–4

		Income Level				
		1	2	3	4	5
Family Size	2					
	3					
	4					
	5					

variable is greatly different from cell to cell the data show *heteroscedasticity*. Overall measures of fit such as $V(e)$ or R^2 are of value in describing results in situations with homoscedasticity (or close to it). In cases of pronounced heteroscedasticity, such overall summaries are deceptive because they overstate the degree of fit in some cells (or for some values of the independent variable or variables) and understate the fit in others.

Regardless of the number of independent variables they use, the simple cross-tabs models (also called one-way ANOVA, or analysis of variance, models) do not assume any pattern in the cell means. The independent variables are used to define the cells, but then the individual cell means are calculated separately with no attempt to force them into any kind of pattern. (In fact, it might be a better illustration to draw Figure 1–4 as a list of 20 different cells with no geometric arrangement, for they are treated simply as a list in the model.) It is natural, however, for analysts to try to see patterns in their results—for example, to see an increase in expenditure with family size and an increase with income. Sometimes models are used to force a pattern on the predicted values produced.

One model that forces a form of pattern is an *additive cross-tabs* model, the simplest version of which has two independent variables. (The model described here is also called a two-way ANOVA without interaction effects.) For example, consider the data in Table 1–12 which might give the number of movies seen in a year for 25 people, each characterized by a level of age and a level of income. For the additive model the prediction portion takes the form

$$a + r_i + c_j$$

where a is a base effect, r_i is the row effect for the ith level or value of one independent variable (called the row variable), and c_j is the column effect for the jth level of the other independent variable (the column variable). In Table 1–12 the row variable is age and the column variable is income. The full equation for this model is

$$y = a + r_i + c_j + e$$

TABLE 1–12

		Income Level				
		1	2	3	4	5
Age-level	1	8	10	12	13	11
	2	2	6	7	11	5
	3	4	10	9	8	10
	4	3	5	9	10	6
	5	9	7	5	5	3

TABLE 1–13

$a = 7.52$	$r_1 = 3.28$	$c_1 = -2.32$
	$r_2 = -1.32$	$c_2 = .08$
	$r_3 = .68$	$c_3 = .88$
	$r_4 = -.92$	$c_4 = 1.88$
	$r_5 = -1.72$	$c_5 = -.52$

y	Predicted Value	e	e^2
8	8.48	− .48	.2304
2	3.88	−1.88	3.5344
4	5.88	−1.88	3.5344
3	4.28	−1.28	1.6384
9	3.48	5.52	30.4704
10	10.88	− .88	.7744
6	6.28	− .28	.0784
10	8.28	1.72	2.9584
5	6.68	−1.68	2.8224
7	5.88	1.12	1.2544
12	11.68	.32	.1024
7	7.08	− .08	.0064
9	9.08	− .08	.0064
9	7.48	1.52	2.3104
5	6.68	−1.68	2.8224
13	12.68	.32	.1024
11	8.08	2.92	8.5264
8	10.08	−2.08	4.3264
10	8.48	1.52	2.3104
5	7.68	−2.68	7.1824
11	10.28	.72	.5184
5	5.68	− .68	.4624
10	7.68	2.32	5.3824
6	6.08	− .08	.0064
3	5.28	−2.28	5.1984
			Sum = 86.56

$V(e) = 3.46$

$V(y) = 8.81$

$$R^2 = \frac{8.81 - 3.46}{8.81} = .61$$

There are various possible ways to define a; one common choice is to make a equal to $M(y)$, the overall mean of the dependent variable. In that case, it turns out that in order to minimize the averaged squared error for this model, the ith row effect r_i will be the difference between $M(y)$ and the mean of the ys for the ith level of the row variable. Each c_j will be the difference between the overall mean $M(y)$ and the mean of the dependent variable for observations with the jth level of the column variable. Table 1–13 shows the predicted values, residuals, and the R^2

value for the data from Table 1–12 when an additive model is fit to them.

The rather extreme case in Table 1–12 suggests another reason for using additive models. There is only one observation per cell—not enough for a simple cross-tabs. While the results in Table 1–13 should not be believed at all strongly, since they are based on so few observations, it was nevertheless possible to fit the additive model. In general, additive models can be considered to substitute model structure for data, as compared with simple cross-tabs. The investigator makes the considerable assumption of additivity—and then can get by with fewer observations than would be needed for simple cross-tabs.

REGRESSION MODELS

A *regression model* with *n* independent variables $x_1, x_2, \ldots, x_n$ is defined by

$$y = b_0 + b_1 * x_1 + b_2 * x_2 + \ldots + b_n * x_n + e$$

Thus, the prediction part of such a model is a linear equation in the *b*s (the coefficients) and in the xs (the independent variables). (One of the xs in this equation may in fact be a nonlinear transformation of some original variable. Hence this regression equation can involve nonlinear relationships between the dependent and an original independent variable—but the equation remains linear in the *b*s). It is often useful to view a regression equation as a way to substitute model structure for data. If we had a huge amount of useful data, then we could consider the cells of a cross-tabulation, one at a time, and consider observations with identical values for the independent variables. For example, to predict expenditures for 4-person high-income families, we would use only data on other high-income 4-person families. Often, however, independent variables take many possible values and we have only a limited number of observations, so that there are not usefully many observations in each cross-tabs cell of interest. We can try to fit a regression equation, substituting a user-supplied form of the equation for the copious data that would be needed for a simpler cross-tabs analysis.

Table 1–14 presents a very small set of data on plant capacity and cost per ton. It shows residuals from the regression equation

$$\text{Cost} = 25.19 - .004404 * \text{capacity}$$

For the data in that table $V(y)$ is 14.46. Hence R^2 for the model is

$$R^2 = \frac{14.46 - 4.35}{14.46} = .70$$

TABLE 1–14

Observation Number	x Plant Capacity (tons/month)	y Average Cost ($/ton)	Predicted Value	e	e^2
1	900	21.95	21.23	.72	.52
2	500	27.18	22.99	4.19	17.56
3	1,750	16.90	17.48	− .58	.34
4	2,000	15.37	16.38	−1.01	1.02
5	1,400	16.03	19.02	−2.99	8.94
6	1,500	18.15	18.58	− .43	.18
7	3,000	14.22	11.98	2.24	5.02
8	1,100	18.72	20.35	−1.63	2.66
9	2,600	15.40	13.74	1.66	2.76
10	1,900	14.69	16.82	−2.13	4.54
				Sum =	43.54
				V(e) =	4.35

(The table contains too few observations to provide sound regression results. As will be true for other examples later in this book, it is included anyway because it is small enough for the reader to see all of the numbers used in measuring the fit of the model.) The coefficients (the *b*s) in a regression equation are found by computer from the input data. In ordinary (least squares) regression the *b*s are selected so as to minimize the resulting average squared error for the model. For example, in the simplest regression equation

$$y = b_0 + b_1 * x + e$$

the computer selects the *slope* b_0 and the *intercept* b_1 that give the smallest $V(e)$ for the input data.

Regression models generally assume *homoscedasticity*—in other words, that the resultant equation will fit as well in one range of values considered for an independent variable as in another such range. In the presence of *heteroscedasticity*, the outputs—especially results on measures of fit—will be misleading, much as they are in cross-tabs.

Especially when they are concerned with structural relationships, investigators often want to try to assign *importance* values to different independent variables in a regression model. In the unlikely case that the independent variables in such an equation are also statistically independent of one another, then such measures are in fact available. In such unlikely cases the coefficient found for variable x_i does not depend on which other variables are included in the equation. Moreover, the decrease in R^2 when variable x_i is dropped from the model does not depend on what other variables are present. Therefore, in such cases this decrease in R^2 could usefully be considered the contribution of x_i and could be used as a measure of x_i's importance.

Usually, however, independent variables are correlated with one another. In such cases, both the coefficient found for a particular variable and the change in R^2 when that variable is dropped from the model depend on which other independent variables are present. There is no really satisfactory solution to the importance problem in such cases.

INTERACTIONS AND ADDITIVITY

In considering *interactions* we focus on the effects of two variables on a third variable. In a model, if the effect of one independent variable (such as age) on a dependent variable (such as dollars spent per month in restaurants) depends on the value of a second independent variable (such as income), we say that there is an interaction between the two independent variables. *Additivity* is the absence of interactions. The cross-tabs models do not assume additivity; the additive cross-tabs and regression models assume additivity.[4]

The issue of interactions or additivity is a separate one from the issue of statistical independence. There can be interactions between independent variables that are statistically independent or statistically dependent; similarly, either statistically independent or statistically dependent variables can have additive effects. (Statistical independence concerns the relation between the two variables themselves. Interaction concerns the relation between their effects on a third variable, the dependent variable.) For example, we might expect age and income to be statistically dependent (in general, older people would earn more than younger ones). If the impact on restaurant spending as income changed were different for younger people than for older ones, there would be an interaction between age and income; if not, the effects of the two variables would be additive.

---

[4]Investigators sometimes include in a regression equation a variable that is the product of two other independent variables. Such a procedure allows the investigator to consider some special types of interaction between variables. In general, however, interactions are included in regression only by conscious effort by the analysts—and only special types of interactions can be considered.

CHAPTER 2

AID (Automatic Interaction Detection)

AID is an exploratory data analysis technique that was developed by researchers at the University of Michigan's Survey Research Center.[1] It is used in considering the relationships between a dependent variable and a number of potential explanatory variables. AID is often an appropriate technique to use when

1. We do not want to make strong assumptions about the relationships.
2. There are more than just a few potential explanatory variables.
3. There are a large number of observations.

The first condition listed often holds in the early stages of an investigation, when we don't feel we understand the data very well. The condition also sometimes holds in situations where we do have some understanding of the data but that understanding makes us believe that assumptions of linearity and additivity are inappropriate. AID is very much like cross-tabulation; neither assumes linearity or additivity.

The second condition listed above is the one that might lead us to avoid cross-tabs and to choose AID. To see this point, consider what would be the process of investigating a set of data with cross-tabs if we had a dependent variable and a set of explanatory variables. With one or two explanatory variables we could easily present the results of a cross-tabulation in a table; each cell of the table would give the mean of the values of the dependent variable on those observations classified in that cell (according to their values on the one or two explanatory variables). With three explanatory variables, we could present cell means in a set of such tables. For example, if we had sex (M or F), age (in one of

[1]See John A. Sonquist, Elizabeth Lauh Baker, and James N. Morgan, *Searching for Structure* (Ann Arbor: University of Michigan Press, 1973).

FIGURE 2–1

Males:

		Age 1 youngest	2	3 oldest
low	1			
income	2			
	3			
high	4			

Females:

		Age 1 youngest	2	3 oldest
low	1			
income	2			
	3			
high	4			

three categories), and income (in one of four categories) as explanatory variables, and if we had yearly expenditures on clothing as the dependent variable, we could set up the tables outlined in Figure 2–1 above. If we filled the cells of these tables with the appropriate cell means, we would have an understandable presentation of relationships.

With this set of tables, some types of comparisons are a bit easier to make than are others. For example, to understand the effects of age, we would compare the cell means in the columns in the male table entry by entry with one another; similarly, we would compare entries in columns in the female table. In this way we would look at the effects of age, with income and sex held constant.

Examining the effects of sex on yearly clothing expenditures is a bit more difficult but could still be accomplished with these tables. For that purpose, we would want to compare the male and female tables entry by entry: the figure for the youngest, lowest-income males with that for the youngest, lowest-income females, and so on.

If we now enlarge the problem and add a fourth explanatory variable, the problem of presenting and understanding the results becomes considerably more difficult. For example, suppose that the additional explanatory variable were occupation, with five possible categories. In order to consider all four explanatory variables at once, we could conceivably construct six tables, one for each age-sex combination, as shown in Figure 2–2.

Even if we had enough observations so that a reasonable number of them fell in each of the cells of these tables, making comparisons with the tables would not be easy. (As an example, consider what would be involved in trying to determine whether the effects of age were the same for males and for females.) As more explanatory variables are added, the problem becomes worse. It becomes almost impossible to present and understand the tables involving all the explanatory varia-

FIGURE 2–2

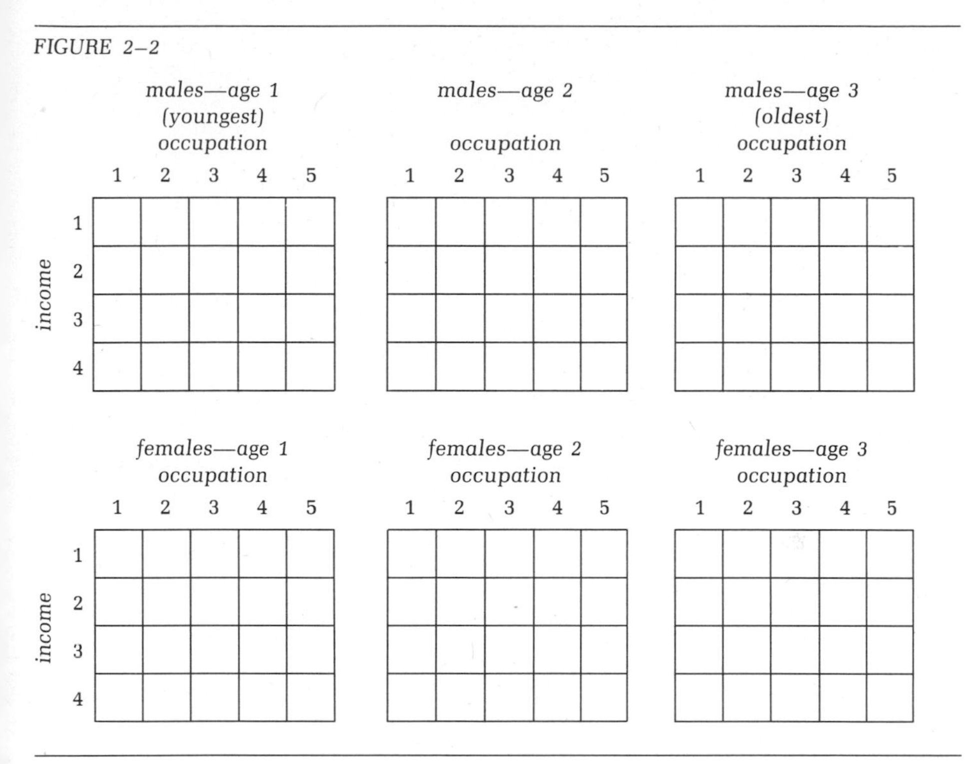

bles. In addition, as we add explanatory variables, the number of cells in the tables grows rapidly and for many databases we soon find many cells with few or no observations.

The usual response of investigators to these problems is to consider the explanatory variables a few at a time with cross-tabs. With 20 potential explanatory variables, they might, for example, consider the explanatory variables three at a time. It may be that the investigators are not sure how the explanatory variables combine and interact and that therefore they want to consider all combinations of three. If so, there will be 1,140 cross-tabs to run. With more explanatory variables, there would be many more cross-tabs. The usual, sensible response of users of cross-tabs is to consider only a subset of the possible cross-tabulations. Using whatever judgment they have about the database, investigators select some of the combinations of explanatory variables and run cross-tabs for those combinations. The results of the first set of cross-tabs may suggest that other combinations be considered, and so on.

The process just described is a laborious one and the results are often hard to present and understand. In addition, the process has another even more serious flaw. Sometimes the effect of one important explanatory variable is seen clearly only if certain other explanatory variables are considered at the same time. An investigator examining a subset of possible tables based on triples of explanatory variables might not include the right triple; indeed, it might take more than three such variables to reveal the effects of one of them.

THE AID PROCEDURE

AID was developed to help facilitate investigations of large databases with many potential explanatory variables. The technique can be thought of as an automated "fishing expedition," using cross-tabs. Rather than having the investigator search laboriously for combinations of explanatory variables that do a good job in explaining the dependent variable, the AID program conducts such a search automatically. In doing so, the AID program does not assume additivity. It does make one special assumption: the explanatory variables useful in explaining one part of the database are not necessarily those most effective for explaining another part. (For example, income might be more important in explaining the yearly clothing expenditures for older individuals in the database than for younger ones.) For further explanation of this point, and of the AID algorithm in general, it is useful to turn to an example.

So that the reader can easily verify all of the steps involved, the following example is a small one, with only 39 observations and 3 potential explanatory variables. Such a problem is far too small for the use of AID for any but these purely illustrative purposes; it does not have enough variables or enough observations to justify real use of AID.

Suppose that we have collected data on 39 adults to try to understand their yearly expenditures on books. For each person in the sample we have collected information on

1. Yearly expenditures on books.
2. Age.
3. Income.
4. Education.

Our sample contains one group of people in their 20s and another group in their 40s; we have coded the age variable 0 for the younger group and 1 for the older one. The income and education variables are also dichotomous. Income is coded 0 for those with dollar incomes below $15,000 and 1 for those with higher incomes. The education variable is 0 for those with a high school education or less and 1 for those with at least some college. The entire database follows in Table 2–1.

The AID algorithm begins by selecting the best division of the full set of data into two subgroups. "Best" is used here in a least-squared-

TABLE 2–1

Observation (Individual #)	*Yearly Expenditures ($)*	Age	Income	Education
1	45	0	0	0
2	35	1	0	0
3	50	1	1	1
4	40	1	1	0
5	48	0	1	0
6	30	1	0	1
7	40	0	0	0
8	55	1	1	1
9	80	0	0	1
10	35	1	0	0
11	48	0	0	0
12	40	0	1	0
13	35	1	0	1
14	47	1	1	0
15	88	0	1	1
16	92	0	0	1
17	52	1	1	1
18	40	1	1	1
19	48	1	1	0
20	85	0	0	1
21	36	1	0	1
22	36	1	0	0
23	98	0	0	1
24	46	1	1	0
25	54	0	0	0
26	55	0	1	0
27	30	1	0	0
28	38	1	0	1
29	90	0	1	1
30	50	0	1	0
31	60	0	1	0
32	30	1	0	1
33	40	1	0	0
34	100	0	0	1
35	50	1	1	0
36	53	1	1	1
37	48	1	1	1
38	34	1	0	0
39	110	0	1	1

error sense. The initial task of the AID algorithm is more precisely stated as follows:

Explaining the dependent variable on each observation in the database by the overall mean gives a figure for the total squared error. Using an explanatory variable to sort observations into two cells and then explaining each observation by the cell mean of its cell will give a lower total squared error. Find that split that gives the lowest resulting total squared error (or, equivalently, find the split that gives the greatest reduction in the total squared error).

For the example problem the overall mean is 53.62. Explaining the dependent variable on each observation by this overall mean gives the results summarized in Table 2–2.

TABLE 2–2

Observation	*Expenditures*	*"Predicted Value"*	*Error*	*Error²*
1	45	53.62	−8.62	74.30
2	35	53.62	−18.62	346.70
.	.	.	.	.
.	.	.	.	.
.	.	.	.	.
39	110	53.62	+56.38	3,178.70
				Total: 18,239

If we classify each observation by age and then proceed to explain the observations by their respective cell means (69.59 for younger and 41.27 for older), the new total squared error of 10,550 can be found as sketched in Table 2–3.

TABLE 2–3

	Observation	*Expenditures*	*"Predicted Value"*	*Error*	*Error²*
Younger	1	45	69.59	−24.59	604.67
	5	48	69.59	−21.59	466.13
	7	40	69.59	−29.59	875.57
	.	.	.	.	.
	.	.	.	.	.
	.	.	.	.	.
	39	110	69.59	40.41	1,632.97
Older	2	35	41.27	−6.27	39.31
	3	50	41.27	+8.73	76.21
	4	40	41.27	−1.27	1.61
	.	.	.	.	.
	.	.	.	.	.
	.	.	.	.	.
	38	34	41.27	−7.27	52.85
				Total:	10,550

If instead we classify each observation by income and then explain the observations by the cell means of 51.05 for lower income and 56.32 for higher, the new total squared error would be 17,969, as sketched in Table 2–4.

TABLE 2–4

	Observation	Expenditures	"Predicted Value"	Error	Error^2
Lower Income	1	45	51.05	−6.05	36.60
	2	35	51.05	−16.05	257.60
	6	30	51.05	−21.05	443.10
	.	.	.	.	.
	.	.	.	.	.
	.	.	.	.	.
	38	34	51.05	−17.05	290.70
Higher Income	3	50	56.32	−6.32	39.94
	4	40	56.32	−16.32	266.34
	.	.	.	.	.
	.	.	.	.	.
	.	.	.	.	.
	39	110	56.32	53.68	2,881.54
				Total:	17,969

Alternatively, we could classify according to education level and then explain with the cell means of 44.05 for lower and 63.68 for higher. The total squared error would be 14,483, as found in Table 2–5.

TABLE 2–5

	Observation	Expenditures	"Predicted Value"	Error	Error^2
Lower Education	1	45	44.05	.95	.90
	2	35	44.05	−9.05	81.90
	4	40	44.05	−4.05	16.40
	.	.	.	.	.
	.	.	.	.	.
	.	.	.	.	.
	38	34	44.05	−10.05	101.00
Higher Education	3	50	63.68	−13.68	187.14
	6	30	63.68	−33.68	1,134.34
	.	.	.	.	.
	.	.	.	.	.
	.	.	.	.	.
	39	110	63.68	46.32	2,145.54
				Total:	14,483

Of the three possible subdivisions of the database just explored, the one resulting in the smallest total squared error is the split on age. (Equivalently, age gives the largest reduction in average squared error: from 18,239 to 10,550.) Therefore, the AID algorithm splits the database into two cells on the basis of age.

It is customary to present AID results in the form of a branching diagram or *AID tree*. The tree in Figure 2–3 shows the results of the initial split: the total sample of 39 observations (N = 39) with mean $\bar{y} = 53.62$ is split into groups 2 and 3 (of sizes N = 17 and N = 22, respectively) with means 69.59 and 41.27.

The *total sum of squares* (TSS) is the sum of the squared errors in group 1, or 18,239; this figure is found by using the overall mean as the predicted value. The *error sum of squares* (ESS) for the first split is the sum of the squared errors for groups 2 and 3, or 10,550; this figure is found by using the group means as predicted values. The difference or reduction in sum of squares is 7,689; this value is also called the *between sum of squares* (BSS) for the split. The ESS is equal to the sum of the squared errors for observations in group 2, called the *within sum of squares* (WSS) for group 2, plus the sum of the squared errors for observations in group 3, called the WSS for group 3. The split into groups 2 and 3 has removed 7,689/18,239 or 42.16 percent of the total sum of squares.[2]

AID is basically a sequential splitting procedure. It proceeds sequentially to perform dichotomous splits, like the one just shown, cre-

FIGURE 2–3

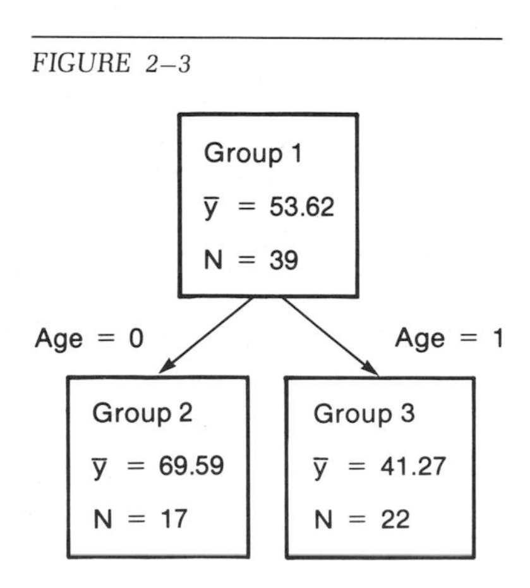

[2]The actual AID computer program provides printed output on the variable selected as the basis of the split, on numbers of observations, on relevant reductions in sums of squares, and so on. Users must translate that output into a diagram such as the tree in Figure 2–3.

ating new subgroups of the data. At any point in the algorithm, AID selects for splitting one of the groups formed on a previous step that has not yet been split. In the AID tree above, groups 2 and 3 satisfy these conditions. They are therefore candidates for splitting. AID selects the candidate group with the largest WSS to be the "parent" group in the next split. In our example, group 2 has a WSS of 9,188, while group 3 has a WSS of 1,362. Therefore, AID tries to split group 2. Two splits are possible:

1. Splitting on income.
2. Splitting on education.

(The group cannot be split on age because it contains only younger people.) Splitting on income would give the tree diagram in Figure 2–4.

FIGURE 2–4

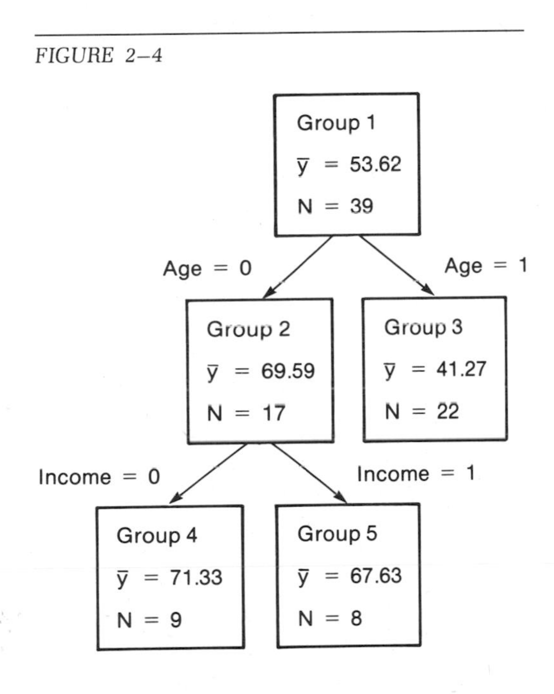

Note that in each such diagram every observation falls into exactly one of the unsplit groups (3, 4, or 5). If each observation is "predicted" by the mean of its unsplit group in this diagram (41.27, 71.33, or 67.63), the sum of squared errors (ESS) turns out to be 10,492. The ESS after the previous split was 10,550. Therefore, this possible split on income reduces the ESS by 58. (58 is the BSS for the split.) This reduction is .32 percent of the TSS of 18,239 (the original value, using the overall mean as predicted value).

If instead we split group 2 on education, we would find the diagram in Figure 2–5.

FIGURE 2–5

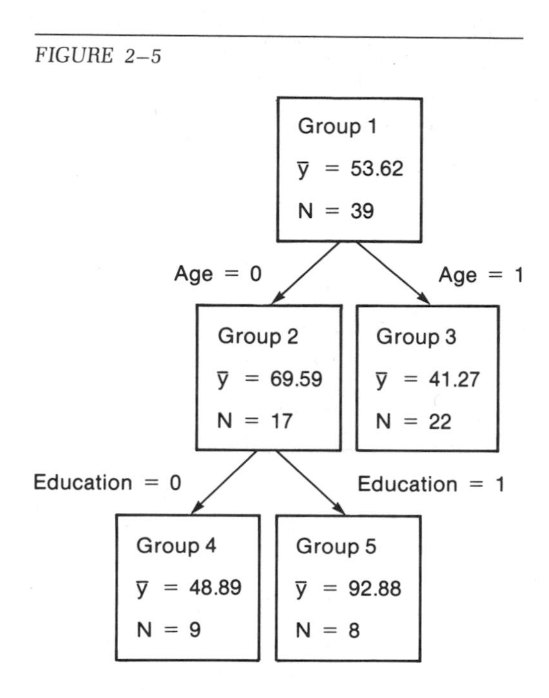

Predicting each observation by the mean of group 3, 4, or 5 in this AID tree gives an ESS of 2,356. The reduction or BSS is 10,550 − 2,356, or 8,194. The BSS is 44.93 percent of the TSS.

Of these two possible splits of group 2, the split on education gives the larger BSS. That split is therefore made.

On the next step, the candidate groups for splitting are 3, 4, and 5. These groups have WSS values of 1,362, 363, and 631, respectively. Group 3 is selected for splitting next because it has the largest WSS.

Group 3 cannot be split on age because it contains only older individuals. If group 3 is split on education, the BSS is 30, or .16 percent of the TSS. If group 3 is split instead on income, the BSS is 1,023 or 5.61 percent of the TSS. (To find these BSS figures, AID determines how much the previous ESS of 2,356 is reduced by predicting each observation in the database by its cell mean after a split of group 3.) The split on income is preferable because it gives the greater BSS. The ESS after the split is 1,333. The splits made so far have explained 16,906 or 92.69 percent of the TSS. The new AID tree is shown in Figure 2–6.

At this point in the process, the statement made at the outset about AID's special assumption should be more clear. Notice that the procedure selected education as the variable to use in explaining the expend-

FIGURE 2–6

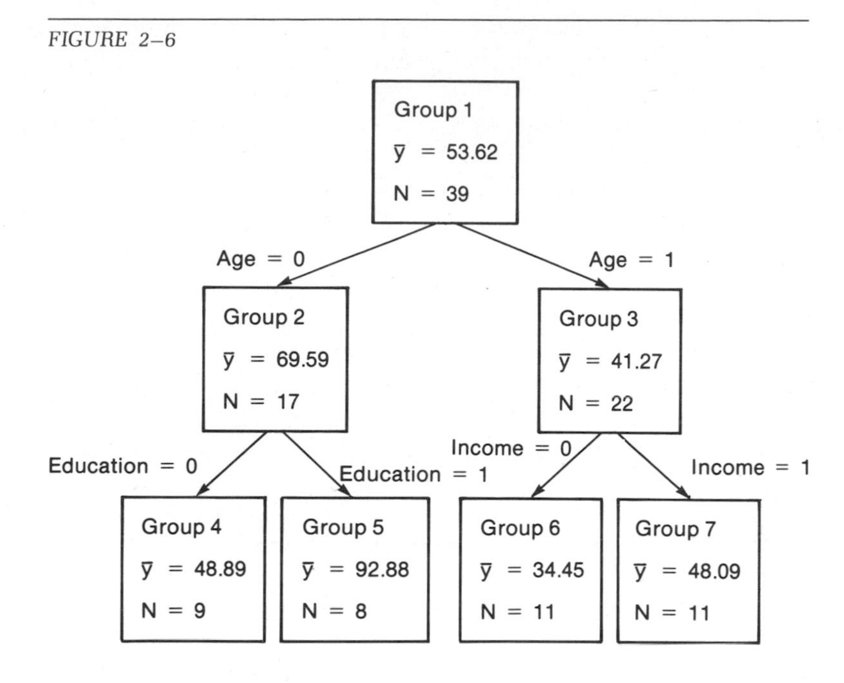

itures of younger individuals but that it selected income for older individuals. Thus, it did not assume that explanatory variables useful in one part of the database were equally useful in explaining another part of the database. In its process, AID has helped us discover an interaction in the effects of the explanatory variables. The effect of education is more pronounced when age is low (younger individuals) than it is when age is high. Similarly, the effect of income is greater for older than for younger individuals. It is this ability of AID to help point out some interactions that prompted its name (Automatic Interaction Detection). The name is in fact not quite accurate, for AID does not automatically detect interactions, nor does it help detect all interactions (as will be discussed below). What AID does do is to allow a form of analysis that does not assume a lack of interactions and that provides investigators with information helpful in identifying some interactions.

The AID procedure continues to make successive splits of groups until it reaches some condition under which it has been instructed to stop. One stopping rule concerns the total number of groups. AID will not create more than some user-specified total number of groups. A second stopping rule concerns the number of observations per group. AID will not create any groups with fewer observations than some other user-specified number. Finally, AID uses a stopping rule based on the explanatory power of its splits. It will not perform any split that reduces the TSS by less than some user-specified percentage. In other words, AID requires some minimum BSS, expressed as a percentage of the TSS. In the current example, the groups that are candidates for further splitting after the first three splits are 4, 5, 6, and 7. If we had specified to AID a minimum explanatory requirement of 1 percent of the TSS, we would find that the algorithm would not split any of these groups; it turns out that no possible split reduces the TSS by this minimum amount. The possible splits are summarized in Table 2–6.

TABLE 2–6

Group to be split	*Variable on which it is split*	*BSS/TSS (percent)*
4	Income	.18%
5	Income	.26
6	Education	.02
7	Education	.18

AID stops without making any of these splits.

INTERPRETING AID RESULTS

The analysis of this sample problem may suggest some tentative conclusions to us. We have found that younger people in the sample spend more on books than do older people. For younger people, those with more education spend considerably more than do those with less. Income is not important in explaining book-buying behavior of the younger group. Some thought about these results may suggest hypotheses. For example, for this younger age group, there may be many students who must spend money on books regardless of whether or not they have much income. It may or may not be true that expenditures for discretionary spending on books (pleasure reading) for such people is tied more closely to income. Our current database does not allow further exploration of the question.

For the older portion of the database the income effect is much stronger than the education effect. This result may suggest that higher income people read more than lower income ones in the older group. On the other hand a totally different conclusion is possible. We have data on expenditures on books, not on reading. It may well be that income does not determine reading but only determines whether people buy books or borrow them from libraries. Perhaps education or some other factors determine reading levels and income merely determines book buying. Again, the database available does not allow exploration of these questions. It has, however, helped raise some interesting issues and it has provided some initial, though tentative, results.

Before we leave this sample problem it can be used to point out a few additional points about AID. First, note that the procedure is a strictly sequential one. At each step it looks for the best available split. It does not start by asking for the best set of four (or some other number of) final groups. We might like to ask for the best set of groups, but it turns out that the computational burden in looking for that set is too great for practicality. In fact, we cannot in general suppose that AID has given us the best set of groups overall; we can only assume that it has made the best choice at each step. For example, suppose that in another problem we have three explanatory variables: socioeconomic status, income, and education. These variables are clearly correlated with one another. It may well be that using income and education gives us final groups that do the best overall job of explaining the database. It may turn out, however, that socioeconomic status is selected for the first split because it is the best single explainer. It may also be true that the other two variables are sufficiently closely related to socioeconomic status that they are not then selected for a split. In such a case AID will provide the best decision at each stage but will *not* provide the best overall set of groups.

Notice in the book-expenditure example that the successive splits did not provide decreasing BSS values. In fact, the second split provided a larger reduction in error than did the first. Such behavior is not uncommon in AID problems. What happened in the example was that the initial split on age provided considerable explanatory power on its own. In addition, it broke out the younger group of individuals, letting the next step provide a further split into two groups (4 and 5) that were quite homogeneous and had small WSS values.

To contrast AID with cross-tabs, it is important to recall that the example used here is far too small for AID for any but illustrative purposes. In this current example we could if we wished have continued to split the groups until no further splits were possible (in other words, until each cell was homogeneous with respect to each of the three explanatory variables). Doing so would have increased the explanatory power from 92.7 percent to 93.3 percent of the TSS. Such steps would give eight cells identical to those that would have been found with a three-way cross-tabs. With many more explanatory variables, proceeding until we had as many cells as there would be in a full cross-tabs would be impractical. In addition, the idea behind AID is to try to make the more powerful splits first. In fact, AID is best suited for databases in which a breakdown into a relatively small number of groups provides substantial explanation of the dependent variable. Hence, proceeding to a full set of groups in such a problem is not only confusing; it also doesn't provide much additional improvement in explanatory power. Thus, the basic idea of AID is to make a sequence of dichotomous splits. The splits can use different explanatory variables in different parts of the tree. We hope that the early splits will provide most of the possible explanation.

VARIABLES IN AID

In applying AID to other problems, there are a few additional points that must be made. In the sample problem above each explanatory variable was coded into only two possible values. In the AID algorithm, up to 32 possible levels are allowed for each explanatory variable.[3] In all cases, however, splits remain dichotomous. For example, even if we had six possible income levels, in splitting on the basis of income AID would define two sets of values. Some of the six levels would be put into one set (or group) and the other levels would be put into the other set. Suppose, for example, that the income levels were as follows:

Level 0. Under $10,000.

Level 1. At least $10,000 but under $12,000.

Level 2. At least $12,000 but under $15,000.

Level 3. At least $15,000 but under $20,000.

Level 4. At least $20,000 but under $25,000.

Level 5. $25,000 and over.

AID might at some point split the group into subgroups on the basis of income, with levels 0, 1, 2, and 3 in one group (group 12, perhaps) and the remaining levels in the other (group 13). Later AID might split group 12 (or one of its descendants) again on income. If it split group 12, it might put income levels 0, 1, and 2 into one subgroup and level 3 into the other. Thus, with variables which have several possible values, each split is dichotomous, but the same variable may be split in different ways at different points in the algorithm.

In the above example of a many-valued variable, we implicitly assumed that the split should be of the form: all levels below some cutoff should go into one subgroup and those above the cutoff should go into the other. In the terminology of AID, variables for which we want to impose such ordering assumptions are called *monotone*. Alternatively, we could have instructed the procedure that any combination of the levels of the income variable into two sets could be used in a split. For example, AID could have put levels 0, 1, 4, and 5 into one subgroup and income levels 2 and 3 into the other. Variables for which we do not impose order conditions are called *free* variables in AID. It should be clear that AID can try many more splits for a free variable than it can for the same variable taken to be monotone.

In the income variable just described, if the user instructs the program to treat the variable as monotone, when AID considers that varia-

[3]Before running AID, users must often recode some of their variables (especially any cardinal variables) into new variables with few levels.

ble for the first time, there are five possible splits (with level 0 in one group; with levels 0 and 1; with levels 0, 1, and 2; with levels 0, 1, 2, and 3; or with levels 0, 1, 2, 3, and 4). If, instead, the variable is assumed free there are 31 possible splits.[4]

Two warnings are in order for the use of free variables. First, as just suggested, there are many possible splits with a free variable. AID is a fishing expedition to begin with and the use of free variables increases considerably the number of choices available. At the same time, free variables increase the likelihood that results will be obtained purely by chance because so many permutations of the data are possible. The second point about free variables is that many variables do in fact have orders associated with them.[5] In many cases, we think it more likely that the income variable should be split into higher and lower income groups and we do not think a group consisting of a mixture of high and low levels (levels 1, 3, and 5 for example) makes sense. In such cases, the income variable should be designated monotone.

On the other hand, in some problems there are variables which should be considered free.[6] For example, there are cases in which it is sensible for middle-income groups to behave differently from those with low or those with high incomes. Or, when we use a variable for geographic region (with different levels for west, mountain, midwest, south, middle Atlantic, and northeast, perhaps) there is not often a natural ordering and we would generally designate the variable as free. Thus, free variables do have a place in AID problems; they should, however, be used with care.

One way to reduce somewhat the potential problems in using a free variable is to include only a few levels for that variable. If the initial coding of the variable has many values, the investigator can judgmentally group those values. For example, a categorical variable for state (in the United States) might be converted to a few-valued categorical variable for region.

This procedure for dealing with free variables suggests a related point for interpreting AID results in general. The more possible parti-

---

[4]The actual AID algorithm uses a shortcut to reduce the computational burden. First, it finds six separate averages of the dependent variable, one for each income level. Next, it sorts the income levels according to increasing values of those averages. Finally, it considers only five possible partitions using the sorted income levels: The partition with only the first income level in one group, the partition with the first two income levels in one group, and so on. This procedure can be proven to give the same best split as would the laborious examination of all 31 potential partitions. Therefore, the problem in using free variables is not primarily one of computational burden. Instead, as explained in the text, it is one of increasing greatly the chances of selecting such a variable (even by chance) and one of difficulty in interpreting the results of such a split.

[5]In particular, ordinal and cardinal variables have natural orders associated with them, and it is often appropriate to require that AID observe those orders.

[6]Categorical variables are by definition unordered, and occasionally it is appropriate to ask AID to disregard the natural ordering of some ordinal or cardinal variable.

tions considered for any variable (whether monotone or free), the more likely it is that variable will be selected on the basis of chance, regardless of any real relationship with the dependent variable. When investigators use different variables with markedly different numbers of levels and allowable partitions, they should bear in mind that a variable with more allowable partitions is, for that reason alone, more likely to be used in a split.

Another Example

The AID tree in Figure 2–7 shows the results of the use of AID to analyze the hours of what is called home production done in 1964 by individuals or couples. (In this example, home production is defined as unpaid work other than regular housework, minus volunteer work and minus courses and lessons.) It appears that family size was a monotone variable; a split was made between six and seven people. Marital status may have been a free variable with three levels (single men, single women, and married couples); the first split grouped all singles in one group and married couples in the other. This AID tree has a rather unusual shape; it is very unsymmetrical. Sonquist, Baker, and Morgan state, "It indicates that any one of several things inhibit such activity or, put another way, only a combination of several favorable factors leads to a substantial amount of such activity. Read the right-hand boxes down the page."[7] (Note that some of the cells have only a few observations. On the other hand the results do seem to make sense.)

OTHER CHARACTERISTICS OF AID

These examples should make clear the general idea behind the AID algorithm. In the words of Morgan and Sonquist in a 1963 paper explaining AID, "The proposal made here is essentially a formalization of what a good researcher does slowly and ineffectively, but insightfully on an IBM sorter. . . . The basic idea is the sequential identification and segregation of subgroups one at a time, nonsymmetrically."[8]

The sequential nature of the algorithm results in some of its properties. For example, the brief discussion above of socioeconomic status, income, and education suggests that the user of AID should be aware of what the procedure can and cannot say about the importance of variables. At each stage, AID selects the single best available explainer. If several variables are correlated, AID may well select one of them and then never choose one of the others. There is some tendency for users to

---

[7]Taken from Sonquist, Baker, and Morgan, *Searching for Structure*, p. 11.

[8]Reprinted in Sonquist, Baker, and Morgan, p. 228.

FIGURE 2–7

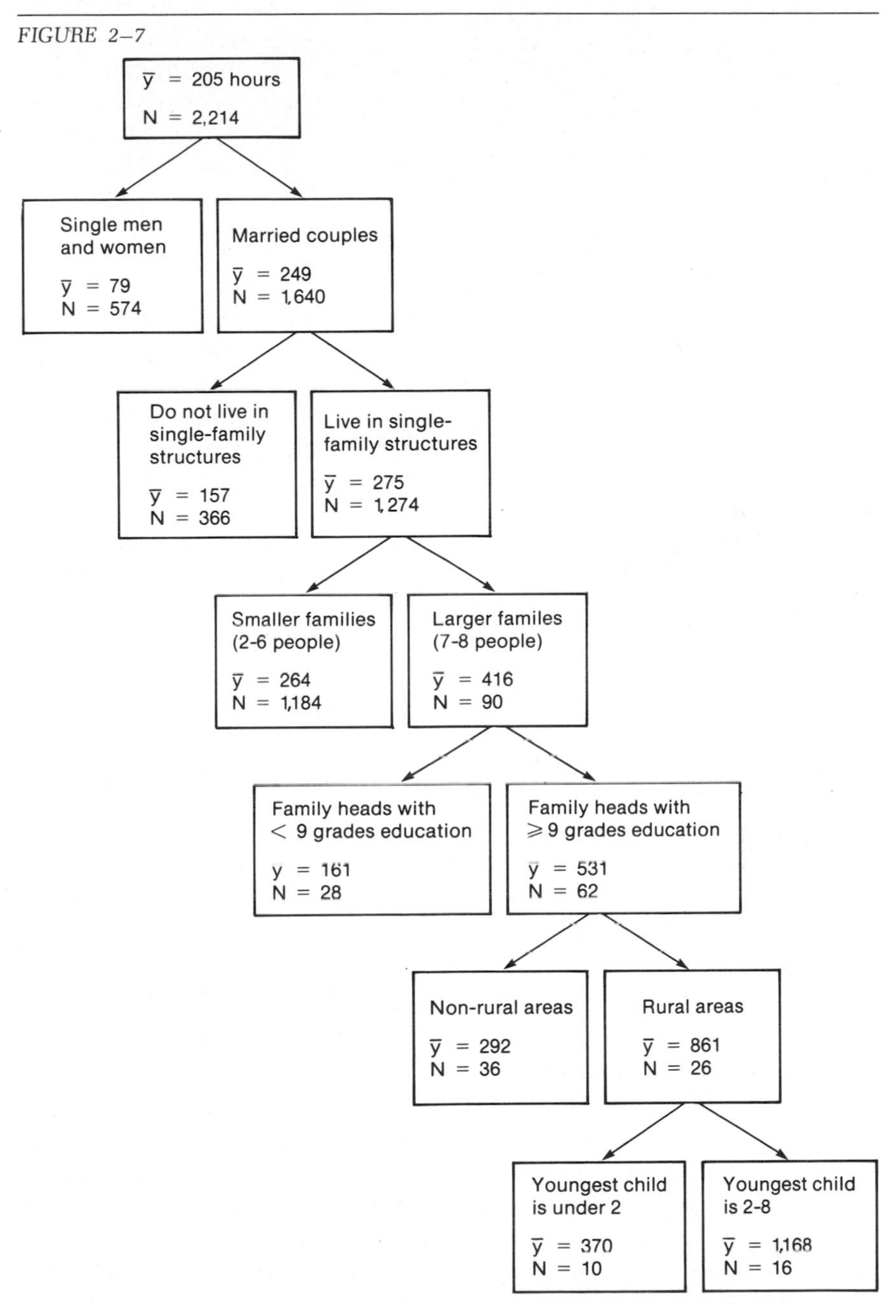

Source: John A. Sonquist, Elizabeth Baker, and James W. Morgan, *Searching for Structure* (Ann Arbor: University of Michigan Press, 1973), p. 12.

want to conclude, as Sonquist, Baker, and Morgan say, that "Perhaps the most striking possible result from the program is the firm conclusion that some particular predictor may not matter."[9]

In fact, however, such conclusions are not really very firm. A variable that does not enter any of the splits in an AID run may in fact be a strong causal factor in the process being investigated; it may not enter simply because it is correlated with other variables that do enter. There is no mechanical way of being sure that the true causal variable (rather than proxies for it) enters the procedure. Moreover, AID's behavior in the face of correlated explanatory variables makes the results of the technique particularly sensitive to small changes (or errors) in the database. Small changes can cause changes in the explanatory variables selected for use. Hence, while AID may provide some evidence about which variables are important, as is true with other procedures (including regression), in the face of correlated variables AID cannot sort out importance.[10]

Some of the optional output provided by AID can help the investigator discover the effects of problems such as collinearity. AID routinely provides summary output on the splits actually chosen, giving the explanatory variable used, the BSS, and so on. (Investigators use this summary output to construct AID trees like the ones shown in this chapter.) As a further option, investigators can request what is called a *trace*. In the trace, the output shows the results of all splits that were explored, not just the ones chosen.[11] If investigators see at some point in the trace that a possible split that was not used gave almost as high a BSS value as the chosen split, they should be warned not to infer that the chosen variable is considerably more powerful than the one not chosen. They may decide that some split that was not actually made but still gave a good BSS is more intuitively sensible than the chosen split. Often, the trace of an AID run shows close decisions of this sort when the explanatory variables are correlated with one another.

Especially because the word *interaction* is part of the name of AID, we should now turn to a consideration of how AID performs in helping the investigator to find interactions. The book-expenditure example has shown AID helping to uncover an interaction. It is worth emphasizing here that AID did not point out the interaction; it did, however, help uncover evidence of the interaction for the investigator to find.

---

[9]Sonquist, Baker, and Morgan, *Searching for Structure*, p. 19.

[10]Notice that regression procedures respond differently in the presence of collinearity. In regression, collinearity affects the values selected for the regression coefficients and increases the uncertainty concerning those values.

[11]In fact, different levels of trace detail may be offered.

There are some types of interactions which AID can help uncover. In other cases, however, the procedure does not help. For example, we will consider two situations. In the first, we might be trying to explain expenditures of couples on a particular type of restaurant. We might find that expenditures were high for higher-income people without children and low for everyone else. The situation is summarized in Figure 2–8.

FIGURE 2–8

	No Children	*Children*
Lower Income	Low	Low
Higher Income	High expenditures	Low

Notice that we have assumed an interaction; the effect of income depends on the presence or absence of children (and vice versa). AID can help find such an interaction. It would first split on one of the two variables and then on the other—since, on average, higher-income couples spend more than lower-income couples, and similarly, on average, those without children spend more than those with children. After two splits AID would have given one subgroup (high income, without children) that was quite unlike the other groups.

For the second example, suppose that we are considering expenditures on clothing, with age (younger or older) and sex (M or F) as the two explanatory variables. Suppose that younger males and older females spend at higher levels and that older males and younger females spend less. The situation is summarized in Figure 2–9.

FIGURE 2–9

	Male	*Female*
Younger	High	Low
Older	Low	High

Note that there is an interaction here; the effect of age depends on sex, and vice versa. The basic AID algorithm may not be able to find this interaction, however. A split on age would give groups with both high spenders and low spenders in each (younger M and F in one; older M and F in the other). It is likely that the resulting group means would be close and that the split would not be considered useful. Similarly, an initial split on sex would result in two groups each of which had high and low spenders. The group means are likely to be similar and the split might not be made. Thus, AID might ignore both variables, because it would only be after two splits that the different groups would be identified. Because AID is a sequential algorithm, it might not proceed far enough to find those groups. The basic AID algorithm helps find only those interactions that are uncovered by a series of single steps. Thus, AID cannot help detect all interactions, although it is useful for many of them.

Finally, we should consider the effect on AID of extreme values of the dependent variable. AID is based on a least squares criterion and, therefore, it is sensitive to extreme values. In some instances, AID will respond to the presence of such values by splitting off small groups containing those values. On the other hand, AID will often not detect outliers. An outlier is a single observation and its effect on its cell mean is diluted by the effects of other observations in the cell. The requirement (or stopping rule) that AID not create groups smaller than some user-specified size increases this dilution. Thus, it is generally appropriate for users to check their databases for errors and extreme values before they run AID. Outliers that do not represent useful data should be eliminated before the AID run.

In summary, AID is a powerful fishing device for use in the preliminary analysis of data. AID's strengths are that it does not require the user to make strong assumptions of additivity, linearity, and the like. Also, it is a powerful automatic search that performs large numbers of calculations efficiently. It can often be used for preliminary analysis of a database that can then be explored further with cross-tabs or other techniques. On the other hand, AID has potential weaknesses that come from the same set of properties. Because it is an automated fishing expedition, AID considers many splits of a database. It may be that the splits it suggests are meaningful in the actual problem but it may also be that the relationships suggested by AID are simply chance variations in the database. The danger is particularly great when free variables are used because so many splits are considered. It is also particularly great when the database is not large (1,000 observations or more are generally needed). The preceding discussion has raised some additional warnings about correlated variables and the importance of variables. Thus, AID is a powerful tool but it must be used with considerable care.

APPENDIX: AID3

A later version of AID, called AID3, adds some additional capabilities to the basic AID operations described in the body of this chapter. This appendix briefly describes some of those capabilities.

The discussion above gave one example of a problem involving interactions in which AID would not help find the interaction because the procedure is so myopic, selecting the best single split only. In AID3 the procedure has been extended so that it looks ahead an additional one or more steps. In other words, on any given step it selects that dichotomous split that gives the best results one or more steps later, in terms of reduction in ESS. The example in the text on clothing expenditures shows one case in which a procedure with look-ahead would succeed in helping find the interaction. One of the splits on sex or on age would turn out to be useful in the problem if the procedure looked ahead an additional step and found that the second split (on the remaining explanatory variable) would be very useful indeed. Again, not all interactions are found with AID3. Interactions can be very complex and even a few steps of look-ahead cannot find all of them. On the other hand, look-ahead does extend the set of interactions that can be found.

The second major change in AID incorporated into AID3 is a fundamental one, involving the definition of the predicted value in a group or cell of the database. In the discussion above, the cell mean was used as the predicted value. AID3 continues to allow the use of the cell mean but it allows other choices as well. In particular, it allows use of regression equations for prediction. For such choices, the algorithm uses a particular variable, called the *covariate*. That variable is the single explanatory variable in the regressions considered. The dependent variable for AID3 is the dependent variable in the regressions. In short, the procedure works as follows. In splitting groups under the regression option in AID3, the procedure makes dichotomous splits on the basis of explanatory variables, as above. After a split, the program finds the regression equation for each group, using the covariate as the explanatory variable in the regression. It then finds residuals and sums of squares of residuals from the group regression equations (rather than residuals from the cell means, as above). It selects the split giving the lowest within-sum-of-squares, as above. It then proceeds with additional splits. For example, suppose that income were the covariate. Suppose that the explanatory variables included age, education, profession, and others. Suppose that the dependent variable were expenditures on books. Then at each stage AID3 would divide one group of observations into two subgroups on the basis of some explanatory vari-

FIGURE 2–10

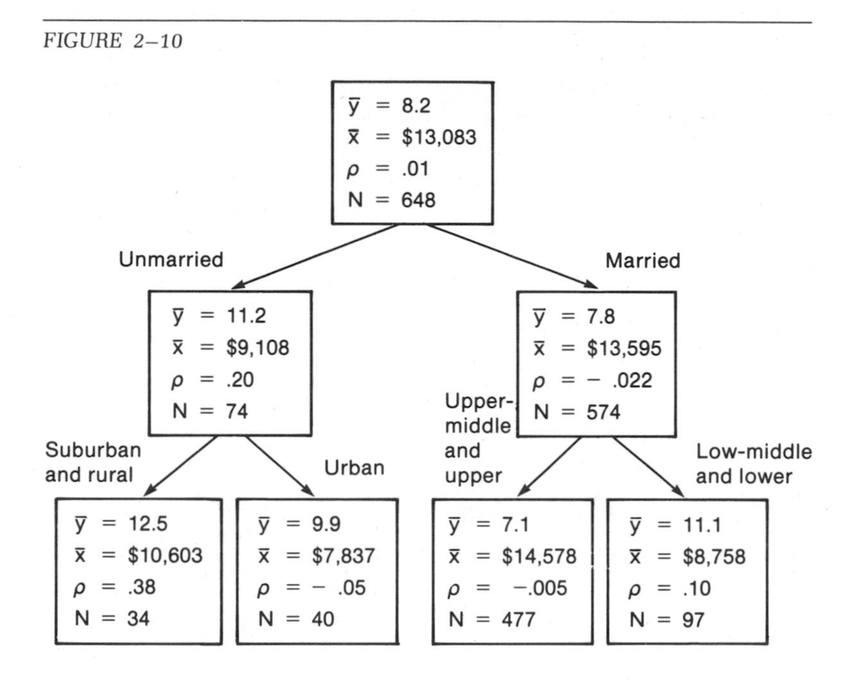

able, such as education. For each unsplit group at that time it would run a separate regression of expenditures on income, and it would determine the resulting residuals to find the ESS (error sum of squares). It would try different splits and select that split that gave the lowest ESS.

The AID3 tree in Figure 2–10 gives an example of a regression analysis in which beer consumption was the dependent variable and income was the covariate.[12] Available explanatory variables were marital status, race, number of children (0, 1–2, 3–5, 6–10), type of area (urban, suburban, rural), home ownership (own or rent), geographic area (east or west), and socioeconomic level (lower, lower-middle, upper-middle, upper). In the boxes, $\bar{y}$ is the mean beer consumption (number of times per month) in the cell, $\bar{x}$ is the mean income, and ρ is the correlation between consumption and income.

In fact, the procedure just described is only one of two regression procedures available in AID3. The following example can motivate the second regression procedure. Suppose that we are working with a wine distributor and we are interested in understanding how people's consumption of wine changes as their income changes. We are not asking about the consumption levels so much as we are asking about marginal

[12]Taken from Vithala R. Rao, John M. McCann, and C. Samuel Craig, "Identifying Market Segments with AID3," *American Marketing Association 1976 Educators Proceedings*, pp. 393–97.

changes in consumption with changes in income. We want to find groups of people who are similar in terms of these marginal changes. For this purpose, we want to find groups where the slopes of the regression of consumption on income are similar; we don't really care whether the intercepts of the regression lines are the same. For example, there may be some people who drink large amounts of wine and others who drink little. If both groups or types of people increase their consumption with income at about the same rate, we may be willing to classify them together. Their levels of consumption differ but their marginal consumptions with income changes do not. Such a procedure might make sense if we were particularly interested in predicting what might happen to consumption in the next few years in an area in which incomes were rising. For cases in which we want to consider the slopes of regression lines only, AID3 allows us to specify such a form of analysis. For such a purpose, it continues to use explanatory variables for dichotomous splits. After a split, it considers regressions, group by group. Within a group, the procedure uses a single value for the regression slope for all observations but it uses different levels (or intercepts) for different subgroups within the group. (Subgroups are identified on the basis of other explanatory variables.) It finds residuals and error-sum-of-squares figures and selects that dichotomous split which gives the biggest reduction in ESS.

These somewhat complicated descriptions should make clear that the regression options of AID3 are considerably more complex than is the basic algorithm. It should also be noted that the regression options involve considerably more assumptions than does the basic algorithm. In particular, they assume linearity in the covariate and they assume that the regression line or the regression slope is the appropriate measure of the relationship to examine in the different cells of the AID tree. Recalling that the motivation for the use of AID was in large part its relatively few assumptions, the reader should see that the more complex options available in AID3 must be used with extreme care.

CHAPTER 3

Conjoint Measurement

Conjoint measurement is a technique closely related in purpose to the additive cross-tabulation (or ANOVA) model and can usefully be considered an extension of ANOVA. Conjoint measurement makes less-restrictive assumptions about the scaling of the dependent variable than does the ANOVA model; as a result, it is particularly well suited for data involving dependent variables such as preference orderings. Recall that the additive cross-tabs model with two independent variables is

$$y = a + r_i + c_j + e$$

Here y is the dependent variable, a is the base, r_i is the effect of the ith level of the row variable, c_j is the effect of the jth level of the column variable, and e is a residual. In this two-way ANOVA model, the dependent variable y is cardinal. In other words, the values of y on different observations can meaningfully be subtracted from one another or divided into one another (and so on) for purposes of comparison. For example, the absolute difference between y values of 14 and 5 and the difference between y values of 19 and 10 are precisely the same. Or, a y value of 4 is exactly twice as great as a value of 2; values of 20 and 10 are similarly related. If you think about a dependent variable giving some individual's ranking of a group of objects from 1 (most desirable) to, perhaps, 25 (least desirable), you will likely find it at least questionable to assume full cardinal properties for the variable y. (Are you sure that the difference between the 1st and 6th ranked objects is exactly the same as the difference between the 11th and 16th?) You may be willing to assume that the effects of two (or more) explanatory variables are in some sense additive, but you would likely not want to assume more than ordinal properties for the dependent variable. Conjoint measurement is intended for such situations. In general, in using the technique one assumes that the independent variables have additive effects but that the dependent variable is only ordinal.[1]

It is also useful by way of introduction to consider the objectives and techniques of conjoint measurement in relation to those of other methods for studying people's preferences for objects. Suppose, for example, that we are analyzing the preferences of individuals for umbrellas and that those preferences depend on quality and on price. One way of studying preferences is to ask separate questions, first about price and then about quality. We might then hope to combine the individual's judgments about quality and price into overall judgments about attractiveness of different price/quality combinations. The process would be called concatenation. Asking questions about price and quality separately will likely not be satisfactory, however. For one thing, everyone will tell us that lower price is always better and that higher quality is also always preferable. What we really want to know is how important the two attributes are relative to one another; in other words, we want to know how individuals trade off price and quality. A similar point is that individuals will often find it very difficult to think of price and quality of umbrellas in isolation, while they will find it quite reasonable to be asked to make judgments about products (umbrellas) which have characteristics along both price and quality dimensions. A major objective of conjoint measurement is to provide information on the trade-offs which individuals make between the various attributes of products or of other objects. In doing so, conjoint measurement can use judgments about entire objects (each characterized by several attributes) rather than direct judgments about the individual attributes of interest.[2] Thus, in an especially common application, conjoint measurement can be thought of as a technique for taking judgments (preference rankings) about objects, each of which is characterized by a set of attributes, and converting that input information into information on the importance an individual assigns to each attribute and the trade-offs s/he would make among them.

EXAMPLE FOR CONJOINT MEASUREMENT

As an example with which to explore the process of conjoint measurement, we will consider data collected from four individuals on their

[1] A more complete explanation is given in the following discussion of what it means to have additive effects on a dependent variable that is only ordinal. In essence, in conjoint measurement the independent variables have additive effects on a monotonic (either non-increasing or non-decreasing) transformation of the dependent variable.

[2] The following discussion will mention a related technique, called trade-off analysis, in which the individual considers only a pair of attributes at any one time.

tastes in bookstores. The questions concerned stores for buying books for pleasure reading rather than school or professional books; each subject was told to think of stores s/he might visit to buy books to read on vacation. The stores were described along four dimensions:

1. Price—the store charged list price or else gave 10 percent off.
2. Parking—the store had its own parking lot or did not.
3. Time—the store took either 5 minutes or 15 minutes for the individual to reach.
4. Selection—the store's selection was placed in one of three categories: Very wide (a large university bookstore, for example), Wide (a more typical bookstore), Medium (one of the smaller bookstores in the suburbs).[3]

Respondents were asked to consider complete descriptions of bookstores along these dimensions—for example,

a. List price, parking, 15 minutes away, wide selection.
b. 10 percent off, parking, 5 minutes away, medium selection.

and so on. A database was set up with the attribute variables coded:

1. 1 for 10 percent off; 2 for list price.
2. 1 for no parking; 2 for parking.
3. 1 for 5 minutes; 2 for 15 minutes.
4. 1 for medium selection; 2 for wide; 3 for very wide.

There was also a preference variable for each person being considered. That variable gave the person's rank-order preference for different stores, with 1 representing the most preferred store. The first table here gives one individual's rankings of 24 store combinations (all possible combinations of the store attributes). The second table contains a different individual's rankings of 12 of the possible 24 store combinations.

CONJOINT MEASUREMENT

We will consider the ratings of one individual at a time. We would like to find a set of effects for price, parking, time, and selection, that can be used to explain Person 1's rank orderings (Table 3–1). Similarly, we would like to find a separate set of effects to explain Person 2's orderings (Table 3–2). Since in conjoint measurement we deal with dependent variables with only ordinal scales, we have to define what it

[3]Respondents were given examples, by name, of each type of store in their area.

TABLE 3–1

Person 1's Preference Ranking	Store Description			
1	10% off	Parking	5 minutes	Very wide
2	List	Parking	5 minutes	Very wide
3	10% off	Parking	15 minutes	Very wide
4	List	Parking	15 minutes	Very wide
5	10% off	Parking	5 minutes	Wide
6	List	Parking	5 minutes	Wide
7	10% off	Parking	15 minutes	Wide
8	List	Parking	15 minutes	Wide
9	10% off	Parking	5 minutes	Medium
10	List	Parking	5 minutes	Medium
11	10% off	Parking	15 minutes	Medium
12	List	Parking	15 minutes	Medium
13	10% off	No parking	5 minutes	Very wide
14	List	No parking	5 minutes	Very wide
15	10% off	No parking	15 minutes	Very wide
16	List	No parking	15 minutes	Very wide
17	10% off	No parking	5 minutes	Wide
18	List	No parking	5 minutes	Wide
19	10% off	No parking	15 minutes	Wide
20	List	No parking	15 minutes	Wide
21	10% off	No parking	5 minutes	Medium
22	List	No parking	5 minutes	Medium
23	10% off	No parking	15 minutes	Medium
24	List	No parking	15 minutes	Medium

TABLE 3–2

Person 2's Preference Ranking	Store Description			
1	10% off	Parking	5 minutes	Very wide
2	List	Parking	5 minutes	Very wide
3	List	Parking	5 minutes	Wide
4	10% off	Parking	5 minutes	Medium
5	10% off	Parking	15 minutes	Wide
6	List	Parking	15 minutes	Wide
7	10% off	Parking	15 minutes	Medium
8	List	No parking	5 minutes	Very wide
9	List	No parking	5 minutes	Wide
10	10% off	No parking	5 minutes	Medium
11	10% off	No parking	15 minutes	Very wide
12	List	No parking	15 minutes	Medium

means to explain such orderings. In this technique, the term *explain* is taken in the following sense:

> We consider a single individual's preference values (which we will call *ys*) for each of the stores. Each store is marked by some level of price, of parking, of time, and of selection. Conjoint measurement is used to find a value for the effect on that individual of each level of price; we might call pr_i the effect of the ith level of price (where i will be 1 or 2). At the same time, conjoint measurement finds values for the effects for the individual of each level of parking, time, and selection. We might call those effects pa_j, t_k, and s_l, respectively.
>
> Now, consider a specific store, marked by the ith level of price, j^{th} level of parking, k^{th} level of time, and l^{th} level of selection. If we add together the effects of each of these attributes for the individual under study, we will obtain a score, which we will call $\hat{Z}$:
>
> $$\hat{Z} = a + pr_i + pa_j + t_k + s_l$$
>
> We want the $\hat{Z}$ values to be close to or explain the original preference values in an ordinal sense. In other words, if y takes a larger value on one observation (store) than on another, we want $\hat{Z}$ to be at least as great for the first store as for the second.[4] We must also make some assumption about what equality of two y values should imply for the corresponding $\hat{Z}$s. The two available choices are (1) to require equality of the $\hat{Z}$s, and (2) not to impose a requirement on the relation of the $\hat{Z}$s. The two choices will be discussed further later. (In our bookstore examples there are no ties in preference, so the problem doesn't arise.)

In practice, it is often not possible to have the $\hat{Z}$ (predicted values) be additive combinations of effects and explain the *ys* in an ordinal sense at the same time. (The following discussion will explain this point further.) For that reason, we introduce residuals (much as we have used residuals in all of the other models we have considered) to measure lack of fit. We assume that there are residuals (*es*) which can be added to the $\hat{Z}$ values to give Z values which are, in fact, explanations of the *ys* in an ordinal sense.

$$Z = \hat{Z} + e$$

In technical terms, the Zs are transformations of the *ys* satisfying the condition that one y that is strictly larger than a second y must correspond to a Z that is at least as large as the Z for the second y. The $\hat{Z}$s are found by addition of the appropriate effects. The Z and $\hat{Z}$ values are as

[4] It might seem preferable to require that the first $\hat{Z}$ be strictly greater than the second. To do so, however, we would have to specify just how much larger would be acceptable (or else the computer could select an extremely small difference). Rather than force an arbitrary choice of the smallest allowable difference, conjoint measurement instead requires only that the first $\hat{Z}$ be at least as great as the second.

close to one another as possible. In measuring the relation between the Zs and the $\hat{Z}$s we use a familiar squared error criterion; we want the sum of the squares of the residuals (es) to be as small as possible.[5] The job of conjoint measurement is to find Zs and additive effects which satisfy these conditions. Notice that this definition involves two types of fit. In considering how well the ys and the Zs fit, we use only an ordinal or ranking criterion and we insist on a perfect fit in that (often very limited) sense. In considering the fit between the Zs and the $\hat{Z}$s we use a squared error criterion, which we try to minimize. In interpreting the results of conjoint measurement, it is important to distinguish between the two types of fit that are involved. The results below will demonstrate this point.

It is useful to contrast the way in which the Zs and the $\hat{Z}$s explain the ys in conjoint measurement with the way in which the effects explain the dependent variable in the ANOVA model. In the additive cross-tabs model, the independent variables are assumed to have additive effects directly on the dependent variable. The differences between the actual values of the dependent variable and the predicted values (or sums of effects) are defined as errors. In conjoint measurement, the independent variables have additive effects on the Zs, which are related to the ys through a transformation. A bit loosely, the transformation Z is called a monotone nondecreasing transformation of the ys; for that reason, conjoint measurement is sometimes called MONotone ANOVA.[6]

FIT IN CONJOINT MEASUREMENT

Before proceeding to some conjoint measurement results for the bookstore example, we now turn to consideration of the relation among ys, Zs, and $\hat{Z}$s in general. In particular, we consider the relationship between the $\hat{Z}$s and the Zs. Suppose that in some hypothetical problem we have identified a possible set of effects to use in explaining (in an ordinal sense) the values of a dependent variable y and that the values of y and of the $\hat{Z}$s are as shown at the top of the next page.

To find values for the residuals in the equation

$$Z = \hat{Z} + e$$

[5]The precise definition of fit involves the sum of the squared es divided by the sum of the squared differences between $\hat{Z}$s and the mean $\hat{Z}$. The definition is discussed further below.

[6]The best-known computer program for conjoint measurement (written by J. B. Kruskal) is called MONANOVA.

y	$\hat{Z}$
1	1.0
2	1.9
3	3.1
4	2.5
5	5.0
6	6.0
7	7.2
8	8.1
9	8.9
10	10.0
11	11.5
12	11.4
13	11.0
14	14.1
15	15.0
.	.
.	.
.	.

we must find values for the Zs which are as close as possible to the $\hat{Z}$s but, at the same time, have the desired ordinal relationship to the ys. For this purpose it is useful to construct a graph plotting the ys against the $\hat{Z}$s. Such a graph is given below in Figure 3–1. The ys are shown along the horizontal axis and the $\hat{Z}$s along the vertical.

FIGURE 3–1

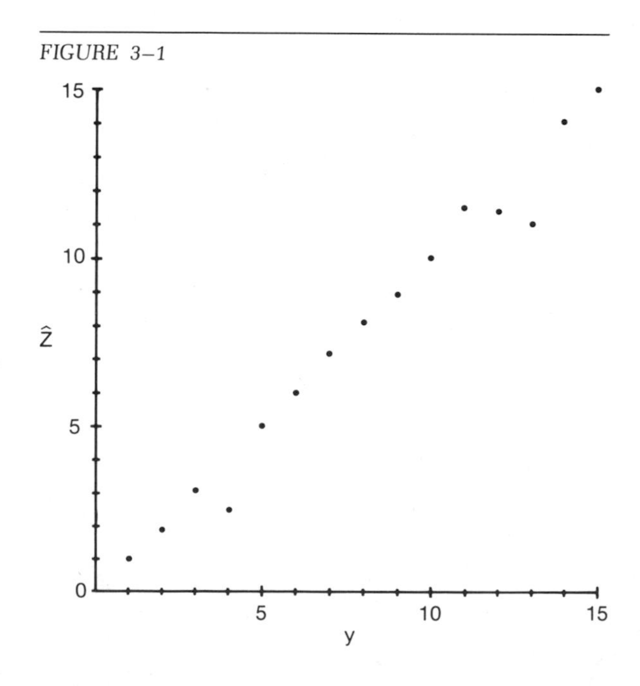

We want to construct a set of Zs which are as close as possible to the $\hat{Z}$s and yet which agree in an ordinal sense with the ys. For the values of Z at y values of 1 and 2 we have no problem; we can let $Z = \hat{Z}$ and satisfy the ordering conditions. For the Zs corresponding to ys of 3 and 4 we have some trouble. The $\hat{Z}$ for $y = 3$ is larger than the $\hat{Z}$ for $y = 4$. The ordering condition requires that the order of the corresponding Zs not be different from that of the ys. At the same time, we want Zs as close as possible to the $\hat{Z}$s. The best solution turns out to be a compromise, with Z equal to 2.8 for $y = 3$ and for $y = 4$. Notice that this choice does in fact satisfy the requirement for ordinal fit between the ys and Zs. You should begin to see why the requirement for ordinal fit is often a rather loose one.

For y values of 5, 6, 7, 8, 9, and 10, the $\hat{Z}$s are in increasing order and we can therefore choose Zs equal to the corresponding $\hat{Z}$s and still fulfill the ordinal fit with y. At $y = 11$ we encounter trouble again. The $\hat{Z}$ at 11 is larger than the value at 12 and that at 13. To make the Zs fit the ys and yet come as close as possible to the $\hat{Z}$s the best choice is a compromise among the three $\hat{Z}$ values. We take Z equal to 11.3 for all three values of y. For $y = 14$ and 15 we can take Z equal to $\hat{Z}$ without violating the ordering requirements. Figure 3–2 shows the Z values on

FIGURE 3–2

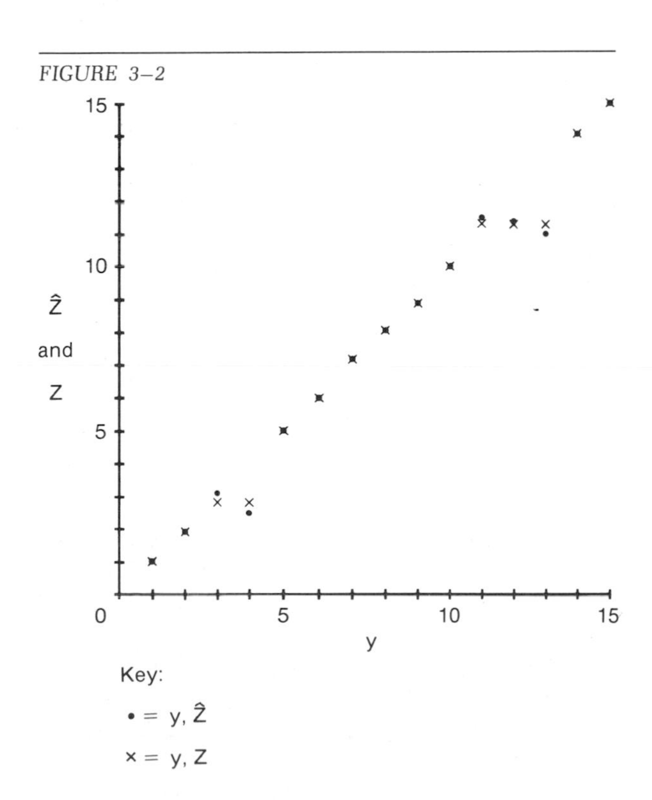

the graph. (The Zs and $\hat{Z}$s coincide at 10 values of *y* and differ at 5 values of *y*.) We can now find the residuals between the Zs and the $\hat{Z}$s; in order of increasing value of *y*, the residuals are 0, 0, −.3, .3, 0, 0, 0, 0, 0, 0, −.2, −.1, +.3, 0, and 0, respectively.

This example shows an important property of conjoint measurement. When we found values of *y* for which the order of the *y*s and the order of the $\hat{Z}$s did not match, we introduced flat or level ranges in our plot of Z, in order to minimize the sum of the squared errors between the Zs and the $\hat{Z}$s. Notice that in such flat areas the $\hat{Z}$s do not tell us about the relative *y* values to which they correspond. In a more extreme case, long flats in the graph of Z against *y* can result in the $\hat{Z}$s' giving very little information indeed about the *y*s; the ordering of the $\hat{Z}$s will imply little about the ordering of the *y*s. Such flat areas on the graphs are called *degeneracy*. It should be clear that degeneracy means that the additive model producing the $\hat{Z}$s does not explain the *y*s very closely in any but the very loose ordinal sense. It should also be clear that we might want to impose some stronger condition on the relation between the *y*s and the Zs than the ordinal ranking condition we have been using. Later in this chapter we will consider the option of imposing some smoothness constraints on this relationship; in other words, we will consider ways of forcing a smoother relationship between *y* and Z.

To understand why these problems of degeneracy might arise, it is important to recall that there are really two steps in conjoint measurement. There is the fit between the additive estimates $\hat{Z}$ and the values of Z, and there is the fit between the Zs and the dependent variable *y*. What the above discussion shows is that the $\hat{Z}$s may fit the Zs well (in the least squares sense) and the Zs may fit the *y*s well (in the ordinal ranking sense) and yet the $\hat{Z}$s may not give much information about the *y*s. In some cases the $\hat{Z}$s do in fact tell a great deal about the *y*s, but in other cases they do not.

OUTPUT OF CONJOINT MEASUREMENT

The output from conjoint measurement takes several forms. First, the graph of the *y*, Z, and $\hat{Z}$ values is generally provided. In addition, the output usually provides the effects that can be used to find the values of the $\hat{Z}$s. As was true for the additive cross-tabulation model, there are different conventions about the base in the listing of effects. The examples below assume that the first level of each variable corresponds to an effect of 0. Therefore, the base value *a* provided is the $\hat{Z}$ value corresponding to having each of the effects at its lowest level. In the bookstore data, this convention means that the base is the $\hat{Z}$ value for a store with 10 percent off in price, no parking, five minutes driving time, and a medium selection. (See the definition of the variables on page 48).

In the terminology of conjoint measurement, the effects of the independent variables are called *part-worths*. To find an estimate of the ranking of an object (store) with some particular configuration of attributes, we add the appropriate part-worths to find a value for $\hat{Z}$. If, as we hope, the $\hat{Z}$ values are close to the Zs, then the $\hat{Z}$ values can be used in an ordinal sense to compare preferences for different objects. In the bookstore example, a higher $\hat{Z}$ corresponds to a less-liked store for the individual under study.

In fact, the part-worths in conjoint measurement tell us a good deal about the individual's preferences. The $\hat{Z}$ scores are found by simple addition. Implicit in them is a concept of compensatory value; changes of a specified magnitude produced by changing the level of one attribute are considered entirely equivalent to changes of the same magnitude produced by changing the level of another attribute. An example should make this statement a bit more clear. Suppose that we are studying umbrellas and that we find the part-worths for four possible price levels to be 0, 3, 4, and 4.8. Suppose that the part-worths for five possible levels of quality are 0, 2, 2.5, 3.3, and 3.7. The base level *a* is 0. We find the $\hat{Z}$ for an umbrella at the 3d level of price and the 4th level of quality as follows:

$$\hat{Z} = 4.0 + 3.3 = 7.3$$

For the 4th level of price and the 3d level of quality the corresponding calculation is

$$\hat{Z} = 4.8 + 2.5 = 7.3$$

The calculations show that for the individual we are considering the change in price from level 3 to level 4 is just balanced by the change in quality from level 4 to level 3 (insofar as the $\hat{Z}$s are good fits to the individual's y or preference values). Similarly, a change in price from level 3 to level 2 would more than offset a change in quality from level 3 to level 4 for that individual. Thus, the part-worths provide information about the trade-offs which the individual would make among attributes.

In addition, we can usefully compare the ranges of values taken by the effects for different attributes. In this example, the range for price is 4.8 while that for quality is 3.7. Therefore, for the set of attribute levels considered, price is a bit more important, in that the maximum relevant change in price can produce a larger change in the value of $\hat{Z}$.

A final important type of output from conjoint measurement describes the fit between the Zs and the $\hat{Z}$s. Recall that a squared-error criterion is used in measuring that fit. The conjoint summary measures include a measure of this error, called the *stress*. It turns out that the most obvious measure, the sum of the squared residuals in the equation

$$Z = \hat{Z} + e$$

is not the best measure.[7] Instead, we really want to know how large the residuals are in relation to the size of the $\hat{Z}$s. Therefore, stress is defined in terms of the sum of the squared residuals divided by the sum of the squared differences between the $\hat{Z}$ values and the mean $\hat{Z}$ value. The stress is the square root of this ratio:

$$\text{Stress} = \sqrt{\frac{\Sigma(Z - \hat{Z})^2}{\Sigma(\hat{Z} - M(\hat{Z}))^2}}$$

Equivalently, the stress is the square root of the ratio of the variance of the residuals divided by the variance of the $\hat{Z}$s. A related measure is the R^2 value for the relation between the Zs and the $\hat{Z}$s. The relationship between Zs and $\hat{Z}$s is an additive cross-tabulation and the R^2 value is defined exactly as it is in additive cross-tabs; it gives the reduction in average squared error when the Z values are predicted by the $\hat{Z}$s rather than by the mean value of Z.

RESULTS OF CONJOINT MEASUREMENT

Table 3–3 and Figure 3–3 give the conjoint measurement results for Person 1's bookstore data. The fits are excellent, both of the *y*s and the

TABLE 3–3

Base	= 21
Price:	0 for 10% off +1 for list
Parking:	0 for no parking −12 for parking
Time:	0 for 5 minutes +2 for 15 minutes
Selection:	0 for medium −4 for wide −8 for very wide
Stress	= .000
R^2	= 1.000

[7]The reason is one of scaling. Suppose we have a candidate set of part-worths, $\hat{Z}$s and Zs. We can calculate the corresponding *e*s. But then, we can always improve the "fit" (reduce the *e*s) by dividing all part-worths, $\hat{Z}$s and Zs by 10, for example. But then, we can divide by 10 again, and so on.

Zs and of the Zs and the $\hat{Z}$s. In fact, this subject's preferences can be predicted exactly by an additive model, as shown in Table 3–4.[8]

Person 1's part-worths show the importance she attaches to the various bookstore attributes. To her, parking is by far the most important factor. Selection is next, then time, and then price. Notice, for example, that she prefers a store with parking but with the least desirable level of each other attribute ($\hat{Z}$ equal 12) to a store with a discount, with five minutes driving time, with a very wide selection but without parking ($\hat{Z}$ equal 13). In numerical terms, the effects for the three other attributes (1, 2, and 8) cannot, combined, make up for the effect of parking (12).

FIGURE 3–3

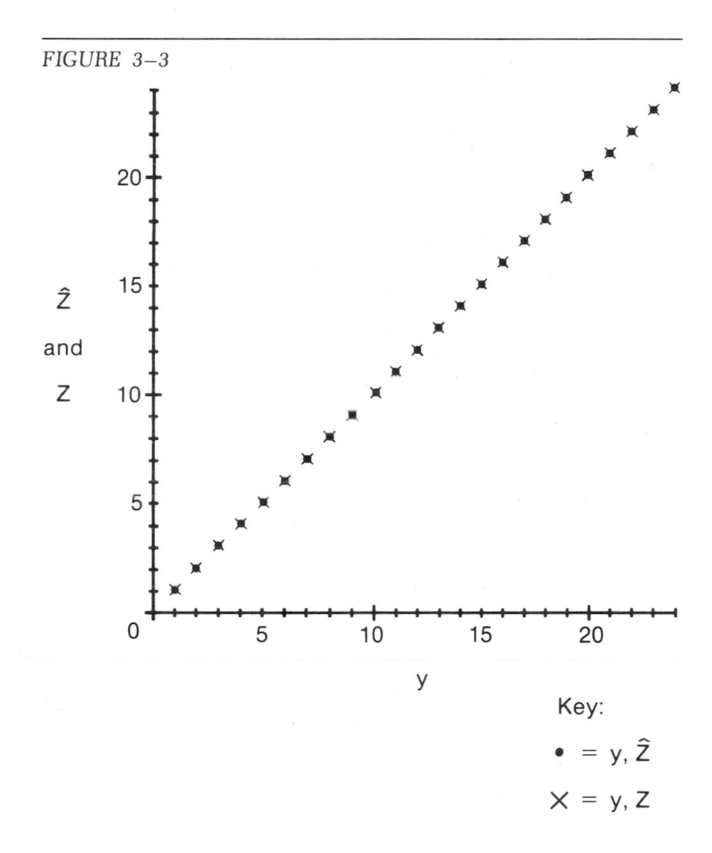

[8]It is worth noting that for Person 1 exactly the same results could have been obtained with a simple (two-way) ANOVA. Conjoint measurement really added nothing to the explanation.

TABLE 3–4

y	$\hat{Z}$	Z
1	1 = 21 + 0 − 12 + 0 − 8	1
2	2 = 21 + 1 − 12 + 0 − 8	2
3	3 = 21 + 0 − 12 + 2 − 8	3
4	4 = 21 + 1 − 12 + 2 − 8	4
5	5 = 21 + 0 − 12 + 0 − 4	5
6	6 = 21 + 1 − 12 + 0 − 4	6
7	7 = 21 + 0 − 12 + 2 − 4	7
8	8 = 21 + 1 − 12 + 2 − 4	8
9	9 = 21 + 0 − 12 + 0 + 0	9
10	10 = 21 + 1 − 12 + 0 + 0	10
11	11 = 21 + 0 − 12 + 2 + 0	11
12	12 = 21 + 1 − 12 + 2 + 0	12
13	13 = 21 + 0 + 0 + 0 − 8	13
14	14 = 21 + 1 + 0 + 0 − 8	14
15	15 = 21 + 0 + 0 + 2 − 8	15
16	16 = 21 + 1 + 0 + 2 − 8	16
17	17 = 21 + 0 + 0 + 0 − 4	17
18	18 = 21 + 1 + 0 + 0 − 4	18
19	19 = 21 + 0 + 0 + 2 − 4	19
20	20 = 21 + 1 + 0 + 2 − 4	20
21	21 = 21 + 0 + 0 + 0 + 0	21
22	22 = 21 + 1 + 0 + 0 + 0	22
23	23 = 21 + 0 + 0 + 2 + 0	23
24	24 = 21 + 1 + 0 + 2 + 0	24

TABLE 3–5

Base	= 9.02
Price:	0 for 10% off +.306 for list
Parking:	0 for no parking −5.5 for parking
Time:	0 for 5 minutes 2.67 for 15 minutes
Selection:	0 for medium −2.29 for wide −2.52 for very wide
Stress	= .000
R^2	= 1.000

Table 3–5 and Figure 3–4 summarize the results for Person 2, when conjoint measurement was applied to her rankings of 12 stores.

These results also are very consistent, although the relation between the Zs and the ys is not one of equality. The Zs and the $\hat{Z}$s match perfectly so that the stress is zero. The ordering of the Zs and that of the ys match, as required, and there are no flats in the graph of Z against y. The fit between her Zs and her $\hat{Z}$s is exact, as shown again in Table 3–6. Person 2 agrees with Person 1 that parking is the most important attribute of bookstores. After that, the two disagree on the ranking of attributes. Person 2 rates time next most important, then selection, and then price. Her part-worths show that for her a change from very wide to wide is more than compensated for by changes from 15 to 5 minutes and from list price to 10 percent off. For Person 1, the opposite was true; the change from very wide to wide was not attractive in exchange for the other two changes.

Next, consider Person 3, whose rankings for bookstores are given in Table 3–7.

FIGURE 3–4

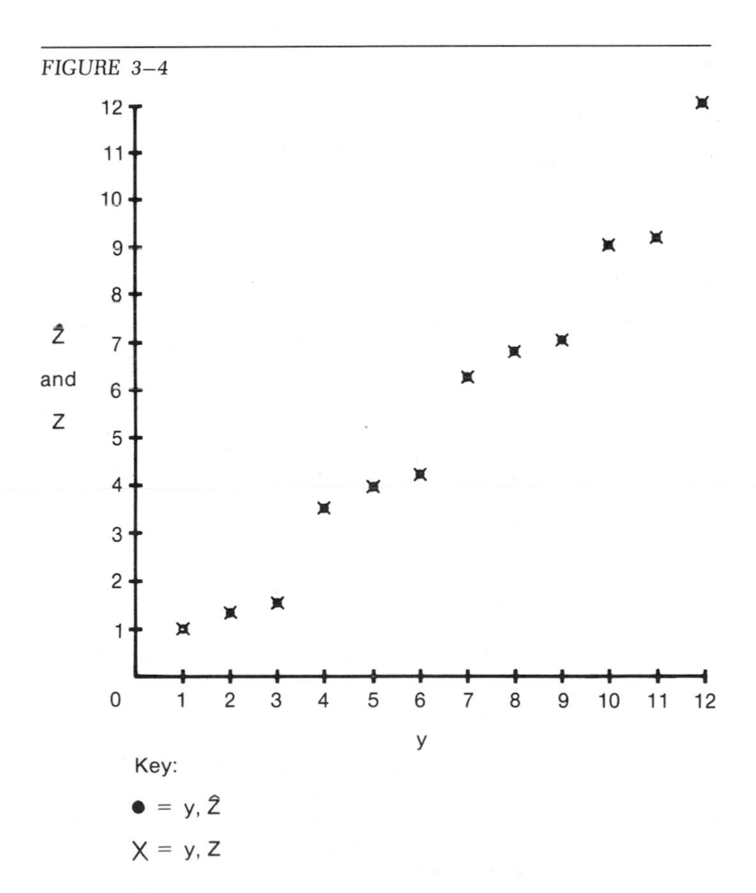

TABLE 3–6

y	$\hat{Z}$	Z
1	$1.000 = 9.02 + 0 - 5.5 + 0 - 2.52$	1.000
2	$1.306 = 9.02 + .306 - 5.5 + 0 - 2.52$	1.306
3	$1.536 = 9.02 + .306 - 5.5 + 0 - 2.29$	1.536
4	$3.520 = 9.02 + 0 - 5.5 + 0 + 0$	3.520
5	$3.900 = 9.02 + 0 - 5.5 + 2.67 - 2.29$	3.900
6	$4.206 = 9.02 + .306 - 5.5 + 2.67 - 2.29$	4.206
7	$6.190 = 9.02 + 0 - 5.5 + 2.67 + 0$	6.190
8	$6.806 = 9.02 + .306 + 0 + 0 - 2.52$	6.806
9	$7.036 = 9.02 + .306 + 0 + 0 - 2.29$	7.036
10	$9.020 = 9.02 + 0 + 0 + 0 + 0$	9.020
11	$9.170 = 9.02 + 0 + 0 + 2.67 - 2.52$	9.170
12	$11.996 = 9.02 + .306 + 0 + 2.67 + 0$	11.996

TABLE 3–7

Person 3's Preference Ranking	Store Description			
1	10% off	Parking	5 minutes	Very wide
2	10% off	Parking	5 minutes	Wide
3	10% off	Parking	5 minutes	Medium
4	List	Parking	5 minutes	Very wide
5	List	Parking	5 minutes	Wide
6	List	Parking	5 minutes	Medium
7	10% off	No parking	5 minutes	Very wide
8	10% off	No parking	5 minutes	Wide
9	10% off	No parking	5 minutes	Medium
10	List	No parking	5 minutes	Very wide
11	List	No parking	5 minutes	Wide
12	List	No parking	5 minutes	Medium
13	10% off	Parking	15 minutes	Very wide
14	10% off	Parking	15 minutes	Wide
15	10% off	No parking	15 minutes	Medium
16	List	Parking	15 minutes	Very wide
17	List	Parking	15 minutes	Wide
18	List	Parking	15 minutes	Medium
19	10% off	No parking	15 minutes	Very wide
20	10% off	No parking	15 minutes	Wide
21	10% off	Parking	15 minutes	Medium
22	List	No parking	15 minutes	Very wide
23	List	No parking	15 minutes	Wide
24	List	No parking	15 minutes	Medium

When subjected to conjoint measurement, Person 3's rankings produce the results shown in Table 3–8 and Figure 3–5.

These results show severe degeneracy. The graph of Z against y consists of two essentially flat regions. Note, however, that in Table 3–9 the fit between Z and $\hat{Z}$ is very good and that the fit between Z and y is correct under the ordinal ranking criterion.

TABLE 3–8

Base	= 1.01
Price:	0 for 10% off +.00392 for list price
Parking:	0 for no parking −.00502 for parking
Time:	0 for 5 minutes +23 for 15 minutes
Selection:	0 for medium −.00137 for wide −.00137 for very wide
Stress	= .000
R^2	= 1.000

FIGURE 3–5

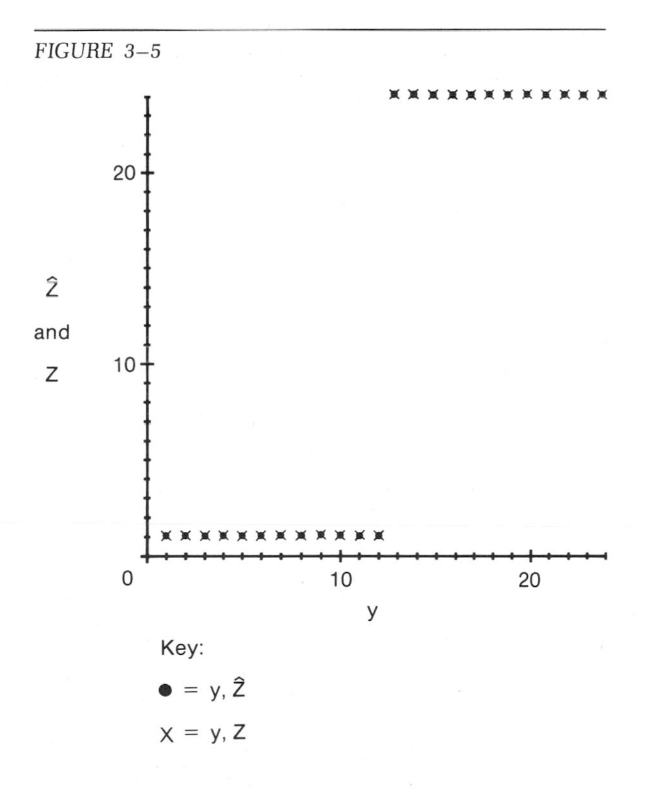

These results clearly do not tell us very much about Person 3's preferences. We might be tempted to try some of the techniques mentioned in passing above for forcing more smoothness on the relation between y and Z. In this case, however, a different approach works. In general, in finding the Z values for a set of additive effects (as shown above), we were forced to use flat regions where the order of the $\hat{Z}$s did

TABLE 3–9

y	$\hat{Z}$	Z
1	1.004 = 1.01 + 0 − .00502 + 0 − .00137	1.004
2	1.004 = 1.01 + 0 − .00502 + 0 − .00137	1.004
3	1.005 = 1.01 + 0 − .00502 + 0 + 0	1.005
4	1.008 = 1.01 + .00392 − .00502 + 0 − .00137	1.008
5	1.008 = 1.01 + .00392 − .00502 + 0 − .00137	1.008
6	1.009 = 1.01 + .00392 − .00502 + 0 + 0	1.009
7	1.009 = 1.01 + 0 + 0 + 0 − .00137	1.009
8	1.009 = 1.01 + 0 + 0 + 0 − .00137	1.009
9	1.010 = 1.01 + 0 + 0 + 0 + 0	1.010
10	1.013 = 1.01 + .00392 + 0 + 0 − .00137	1.013
11	1.013 = 1.01 + .00392 + 0 + 0 − .00137	1.013
12	1.014 = 1.01 + .00392 + 0 + 0 + 0	1.014
13	24.004 = 1.01 + 0 − .00502 + 23 − .00137	24.004
14	24.004 = 1.01 + 0 − .00502 + 23 − .00137	24.004
15	24.005 = 1.01 + 0 + 0 + 23 + 0	24.005
16	24.008 = 1.01 + .00392 − .00502 + 23 − .00137	24.008
17	24.008 = 1.01 + .00392 − .00502 + 23 − .00137	24.008
18	24.009 = 1.01 + .00392 − .00502 + 23 + 0	24.009
19	24.009 = 1.01 + 0 + 0 + 23 − .00137	24.009
20	24.009 = 1.01 + 0 + 0 + 23 − .00137	24.009
21	24.010 = 1.01 + 0 − .00502 + 23 + 0	24.010
22	24.013 = 1.01 + .00392 + 0 + 23 − .00137	24.013
23	24.013 = 1.01 + .00392 + 0 + 23 − .00137	24.013
24	24.014 = 1.01 + .00392 + 0 + 23 + 0	24.014

TABLE 3–10

Base	= 9.0
Price:	0 for 10% off +3 for list
Parking:	0 for no parking −6 for parking
Time:	0 for 5 minutes 12 for 15 minutes
Selection:	0 for medium −1 for wide −2 for very wide
Stress	= .000
R^2	= 1.000

FIGURE 3–6

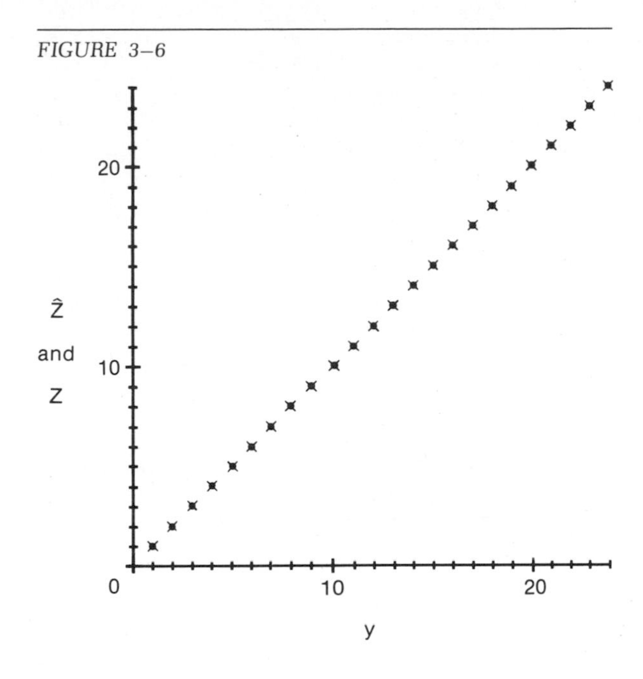

not match the ordering of the ys. One possible reason for this phenomenon is the presence in the data of outliers or errors. Looking at the original table of Person 3's ranking suggests the presence of such an error. If we switch two stores in his ordering (the 10 percent off, no parking, 15 minutes, medium store with the 10 percent off, parking, 15 minutes, medium store) and rerun the conjoint program, we find the results shown in Table 3–10 and Figure 3–6. Person 3's data now provide very sensible results. He values time most, then parking, then price, and then selection. In fact, his preferences can now be explained exactly as an additive model, as shown in Table 3–11, where the $\hat{Z}$ values exactly equal the ys. It would not be appropriate to change a person's rankings without checking with him or her to see if we had in fact identified an error. In the current example, Person 3 felt that he had made an error in his original ranking and that the analysis should be based on the corrected list.

Finally, consider another set of rankings, from Person 4. The input data are summarized in Table 3–12. The conjoint measurement results are given in Table 3–13 and Figure 3–7. This person's data also show degeneracy. If we examine her input judgments we find that the problem is not as simple as that for Person 3. For example, we notice that her 7th and 5th rankings suggest that when parking is available she prefers a reduction in time to a change from very wide to wide. Yet, her 9th and 11th rankings suggest that where there is no parking she prefers

TABLE 3–11

y	$\hat{Z}$	Z
1	1 = 9 + 0 − 6 + 0 − 2	1
2	2 = 9 + 0 − 6 + 0 − 1	2
3	3 = 9 + 0 − 6 + 0 + 0	3
4	4 = 9 + 3 − 6 + 0 − 2	4
5	5 = 9 + 3 − 6 + 0 − 1	5
6	6 = 9 + 3 − 6 + 0 + 0	6
7	7 = 9 + 0 + 0 + 0 − 2	7
8	8 = 9 + 0 + 0 + 0 − 1	8
9	9 = 9 + 0 + 0 + 0 + 0	9
10	10 = 9 + 3 + 0 + 0 − 2	10
11	11 = 9 + 3 + 0 + 0 − 1	11
12	12 = 9 + 3 + 0 + 0 + 0	12
13	13 = 9 + 0 − 6 + 12 − 2	13
14	14 = 9 + 0 − 6 + 12 − 1	14
15	15 = 9 + 0 − 6 + 12 + 0	15
16	16 = 9 + 3 − 6 + 12 − 2	16
17	17 = 9 + 3 − 6 + 12 − 1	17
18	18 = 9 + 3 − 6 + 12 + 0	18
19	19 = 9 + 0 + 0 + 12 − 2	19
20	20 = 9 + 0 + 0 + 12 − 1	20
21	21 = 9 + 0 + 0 + 12 + 0	21
22	22 = 9 + 3 + 0 + 12 − 2	22
23	23 = 9 + 3 + 0 + 12 − 1	23
24	24 = 9 + 3 + 0 + 12 + 0	24

TABLE 3–12

Person 4's Preference Ranking	*Store Description*			
1	10% off	Parking	5 minutes	Very wide
2	List	Parking	5 minutes	Very wide
3	10% off	No parking	5 minutes	Very wide
4	List	No parking	5 minutes	Very wide
5	10% off	Parking	5 minutes	Wide
6	List	Parking	5 minutes	Wide
7	10% off	Parking	15 minutes	Very wide
8	List	Parking	15 minutes	Very wide
9	10% off	No parking	15 minutes	Very wide
10	List	No parking	15 minutes	Very wide
11	10% off	No parking	5 minutes	Wide
12	List	No parking	5 minutes	Wide
13	10% off	Parking	15 minutes	Wide
14	List	Parking	15 minutes	Wide
15	10% off	No parking	15 minutes	Wide
16	List	No parking	15 minutes	Wide
17	10% off	Parking	15 minutes	Medium
18	List	Parking	15 minutes	Medium
19	10% off	No parking	5 minutes	Medium
20	List	No parking	5 minutes	Medium
21	10% off	Parking	5 minutes	Medium
22	List	Parking	5 minutes	Medium
23	10% off	No parking	15 minutes	Medium
24	List	No parking	15 minutes	Medium

TABLE 3–13

Base	= 24
Price:	0 for 10% off −.00000103 for list
Parking:	0 for no parking −.00000537 for parking
Time:	0 for 5 minutes +.00000746 for 15 minutes
Selection:	0 for medium −23 for wide −23 for very wide
Stress	= .000
R^2	= 1.000

a change from very wide to wide to a change from 15 to 5 minutes. For this individual, the conjoint model does not fit well at all. There seem to be interactions in her responses.[9] In other words, the effect of change in level for one variable, such as time, seems to depend on the level of other variables, such as parking. The basic conjoint measurement algorithm does not allow for such interactions and it will not give good fits to Person 4's responses.

FIGURE 3–7

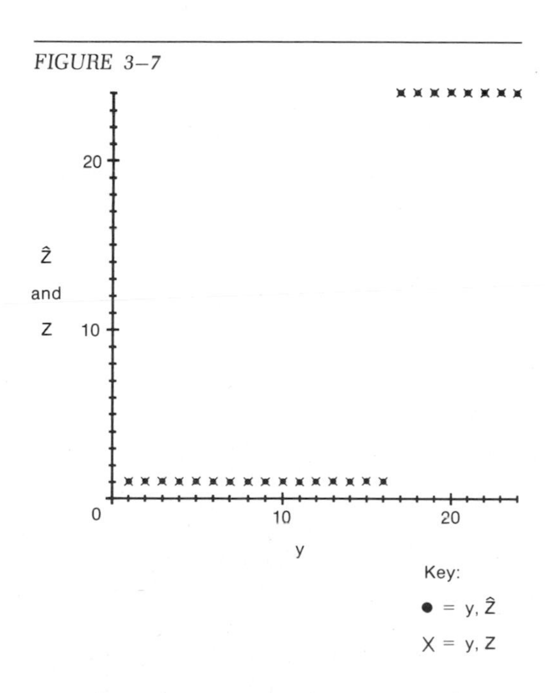

[9]Either that or else she didn't understand the instructions for ranking the stores.

ADDITIONAL ISSUES

The actual procedure used in conjoint measurement to find a set of part-worths and a set of values for Z involves a search. The computer program starts with some set of values as candidates for the effects of attributes. It then finds the best set of Zs, much as we did above. The Zs must agree with the *y*s in an ordinal sense and must be as close as possible to the $\hat{Z}$s. The program then calculates the stress. Unless that stress value is zero or very small, the program proceeds to search. It selects another set of candidate part-worths (by a numerical procedure for finding part-worths that would reduce the error sum of squares). It then finds a new set of $\hat{Z}$s and Zs and calculates another value for the stress. If those values are good enough the procedure stops. If not, it continues to search. Programs for conjoint measurement will in general have limitations on the number of search steps or iterations that can be taken. They will also have a value for the minimum acceptable improvement in the stress per iteration. Thus, a program may stop if it has reached the lowest possible stress (or even zero stress), if it has used the maximum allowable number of iterations, or if the stress levels are not being appreciably reduced by additional iterations.

None of the bookstore examples considered above contained ties. In other words, the subjects assigned the bookstores a strict preference ranking. In other cases, conjoint measurement is used on data which can include ties. As mentioned briefly above, there are two ways of handling ties. In one approach, we insist that equality of two *y* values be reflected in equality of the corresponding Zs. The implications of this choice in the graphical procedure for finding Zs from the *y*s should be clear. The alternative choice is to assume that equality of the *y*s does not necessarily imply equality of the Zs. This second choice is often attractive in situations involving preference data. Suppose individuals are asked to assign preference scores on a scale of 1 to 10 to a group of objects. We may want to assume that ties in scores may not really mean that an individual finds the two objects equally attractive. Instead, a tie may mean that the objects are close in attractiveness and that the individual has not taken the time or trouble to differentiate further between them. If so, we may want to treat ties in the second way. It is important, however, to realize that making the second choice for treating ties reduces even further the restriction placed on the relationship between the *y*s and the Zs. We saw above that the looseness of the correspondence between *y*s and Zs could limit considerably the usefulness of results from conjoint measurement. Users should be careful, in assuming that ties are not meaningful, to be aware of the increased potential that the *y*s and Zs will not correspond closely to one another in other than a very loose ordinal sense.

FORCING SMOOTHNESS

This discussion should suggest further the desirability in some situations of imposing stricter conditions on the relationship between the ys and the Zs. One possibility for doing so is to require that the relationship be smooth. One way to make the definition of smoothness operational is to require that the Zs be a polynomial function of y. In other words, we require that

$$Z = a_0 + a_1 y + a_2 y^2 + \ldots a_n y^n$$

for a polynomial of some degree n. In addition, we want the Zs as close as possible, in a least squares sense, to the additive estimates given by the $\hat{Z}$s. For example, consider the following set of data. The managers of the cafeteria on a university campus are interested in determining the preferences of their customers for meals characterized by different size portions and by different prices. There are three possible levels of portion (relative to the other portions offered at a particular meal) and there are five possible levels of relative price. The managers have asked customers to assign ratings, on a scale from 0 to 50, to combinations of price and portion. On this scale, 50 is the best rating. Data from one customer are given in Table 3–14.

These data were subjected to conjoint analysis with a polynomial of degree 2. (The degree of the polynomial to be used is a choice made by the user of the program.) The results are summarized in Table 3–15 and Figure 3–8.

TABLE 3–14

Portion	*Price*	*Rating*
0 = Smaller	1 = Much higher	
1 = Equal	2 = Somewhat higher	
2 = Larger	3 = Equal	
	4 = Somewhat lower	
	5 = Much lower	
0	1	10
0	2	21
1	1	27
0	4	33
0	5	34
2	1	36
1	3	37
1	4	37
2	2	40
2	3	40
1	5	43
2	5	45

TABLE 3–15	
Base	= 9.96
Portion:	0 for level 0
	10.7 for level 1
	18.5 for level 2
Price:	0 for level 1
	5.94 for level 2
	9.22 for 3
	13.4 for 4
	18.1 for 5
Stress	= .178
R^2	= .968

FIGURE 3–8

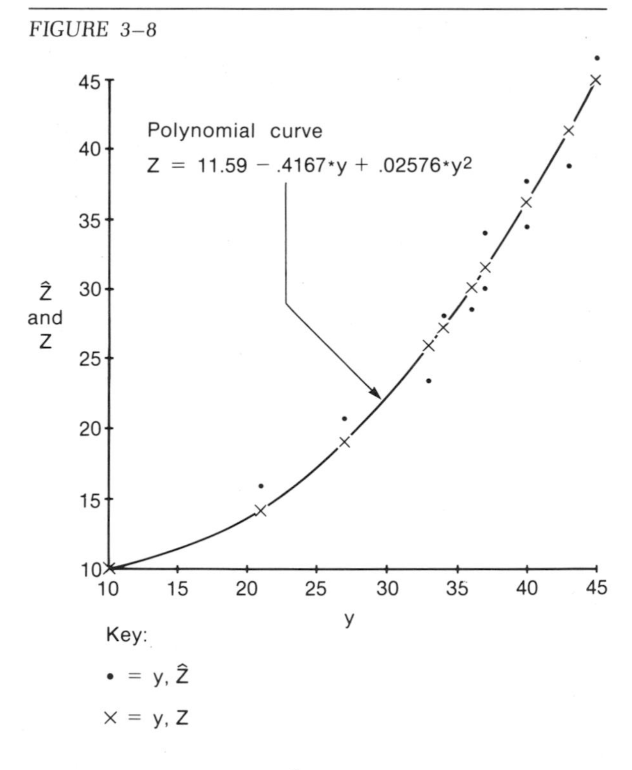

Table 3–16 gives the values of y, $\hat{Z}$, and Z. Notice that the results of this fit are smooth and that the additive model fits the Zs well.

In the price and portion example the data give a set of Z values which are monotone as well as smooth. Unfortunately, the monotonicity of such results cannot be guaranteed. In other words, in using polynomial fits, the conjoint program may give sets of Z values that increase and then decrease as shown in Figure 3–9.

TABLE 3–16

y	$\hat{Z}$	Z*
10	9.96 = 9.96 + 0 + 0	10.0
21	15.90 = 9.96 + 0 + 5.94	14.2
27	20.66 = 9.96 + 10.7 + 0	19.1
33	23.36 = 9.96 + 0 + 13.4	25.9
34	28.06 = 9.96 + 0 + 18.1	27.2
36	28.46 = 9.96 + 18.5 + 0	30.0
37	29.88 = 9.96 + 10.7 + 9.22	31.4
37	34.06 = 9.96 + 10.7 + 13.4	31.4
40	34.40 = 9.96 + 18.5 + 5.94	36.1
40	37.68 = 9.96 + 18.5 + 9.22	36.1
43	38.76 = 9.96 + 10.7 + 18.1	41.3
45	46.56 = 9.96 + 18.5 + 18.1	45.0

*Z is $11.59 - .4167 * y + .02576 * y^2$.

With higher order polynomials, there can be several reversals in directions of the graph of Z. It is not possible to require that the graph of Z against y be both monotone nondecreasing and smooth as well. The polynomial option is attractive in that it imposes a much stricter relation between the ys and the Zs than does the more general ordinal relation, but neither choice is fully satisfactory.

FIGURE 3–9

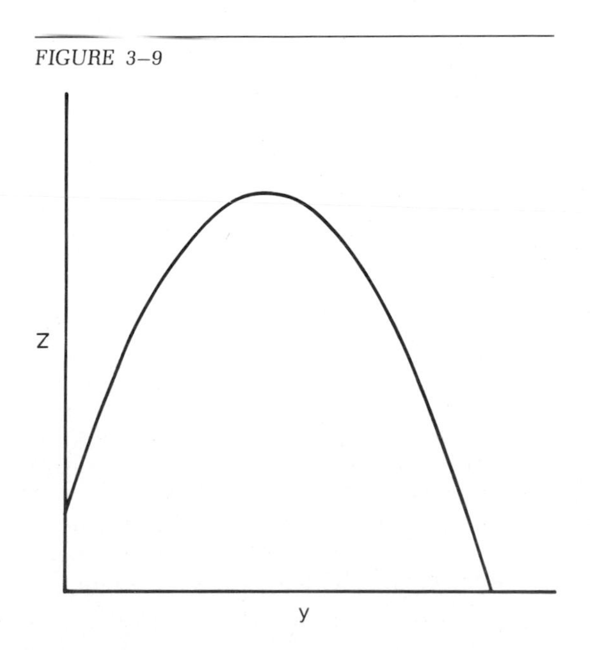

ANALYSIS FOR MORE THAN ONE INDIVIDUAL

All of the results discussed so far have considered one individual at a time. It is fairly common in applications of conjoint measurement to find investigators combining data for several individuals. Presumably, their motivation is to try to summarize the results for a whole set of individuals in a more concise form. Yet, doing so is often very dangerous, in that the methods commonly used obscure the actual data. Frequently, investigators obtain average or composite ranks for a group of individuals by averaging the individual rankings. In Table 3–17 the rankings for Person 1 and Person 3 are repeated from above. Rankings given to the full set of 24 store combinations by Person 2 are also given, as is the average of the three rankings. Table 3–18 and Figure 3–10 show the results of applying conjoint measurement to the average rankings.

TABLE 3–17

Store Description				*Person 1's Ranking*	*Person 2's Ranking*	*Person 3's Ranking*	*Average Ranking*
10% off	Parking	5 minutes	Very wide	1	1	1	1
10% off	Parking	5 minutes	Wide	5	3	2	3⅓
10% off	Parking	5 minutes	Medium	9	5	3	5⅔
10% off	Parking	15 minutes	Very wide	3	7	13	7⅔
10% off	Parking	15 minutes	Wide	7	9	14	10
10% off	Parking	15 minutes	Medium	11	11	15	12⅓
10% off	No parking	5 minutes	Very wide	13	13	7	11
10% off	No parking	5 minutes	Wide	17	15	8	13⅓
10% off	No parking	5 minutes	Medium	21	17	9	15⅔
10% off	No parking	15 minutes	Very wide	15	19	19	17⅔
10% off	No parking	15 minutes	Wide	19	21	20	20
10% off	No parking	15 minutes	Medium	23	23	21	22⅓
List	Parking	5 minutes	Very wide	2	2	4	2⅔
List	Parking	5 minutes	Wide	6	4	5	5
List	Parking	5 minutes	Medium	10	6	6	7⅓
List	Parking	15 minutes	Very wide	4	8	16	9⅓
List	Parking	15 minutes	Wide	8	10	17	11⅔
List	Parking	15 minutes	Medium	12	12	18	14
List	No parking	5 minutes	Very wide	14	14	10	12⅔
List	No parking	5 minutes	Wide	18	16	11	15
List	No parking	5 minutes	Medium	22	18	12	17⅓
List	No parking	15 minutes	Very wide	16	20	22	19⅓
List	No parking	15 minutes	Wide	20	22	23	21⅔
List	No parking	15 minutes	Medium	24	24	24	24

TABLE 3–18

Base	= 15.67	Time:	0 for 5 minutes +6.67 for 15 minutes
Price:	0 for 10% off +1.67 for list	Selection:	0 for medium −2.33 for wide −4.67 for very wide
Parking:	0 for no parking −10 for parking		

Stress = .000
R^2 = 1.000

y	$\hat{Z}$	Z
1	1 = 15⅔ + 0 − 10 + 0 − 4⅔	1
2⅔	2⅔ = 15⅔ + 1⅔ − 10 + 0 − 4⅔	2⅔
3⅓	3⅓ = 15⅔ + 0 − 10 + 0 − 2⅓	3⅓
5	5 = 15⅔ + 1⅔ − 10 + 0 − 2⅓	5
5⅔	5⅔ = 15⅔ + 0 − 10 + 0 + 0	5⅔
7⅓	7⅓ = 15⅔ + 1⅔ − 10 + 0 + 0	7⅓
7⅔	7⅔ = 15⅔ + 0 − 10 + 6⅔ − 4⅔	7⅔
9⅓	9⅓ = 15⅔ + 1⅔ − 10 + 6⅔ − 4⅔	9⅓
10	10 = 15⅔ + 0 − 10 + 6⅔ − 2⅓	10
11	11 = 15⅔ + 0 + 0 + 0 − 4⅔	11
11⅔	11⅔ = 15⅔ + 1⅔ − 10 + 6⅔ − 2⅓	11⅔
12⅓	12⅓ = 15⅔ + 0 − 10 + 6⅔ + 0	12⅓
12⅔	12⅔ = 15⅔ + 1⅔ + 0 + 0 − 4⅔	12⅔
13⅓	13⅓ = 15⅔ + 0 + 0 + 0 − 2⅓	13⅓
14	14 = 15⅔ + 1⅔ − 10 + 6⅔ + 0	14
15	15 = 15⅔ + 1⅔ + 0 + 0 − 2⅓	15
15⅔	15⅔ = 15⅔ + 0 + 0 + 0 + 0	15⅔
17⅓	17⅓ = 15⅔ + 1⅔ + 0 + 0 + 0	17⅓
17⅔	17⅔ = 15⅔ + 0 + 0 + 6⅔ − 4⅔	17⅔
19⅓	19⅓ = 15⅔ + 1⅔ + 0 + 6⅔ − 4⅔	19⅓
20	20 = 15⅔ + 0 + 0 + 6⅔ − 2⅓	20
21⅔	21⅔ = 15⅔ + 1⅔ + 0 + 6⅔ − 2⅓	21⅔
22⅓	22⅓ = 15⅔ + 0 + 0 + 6⅔ + 0	22⅓
24	24 = 15⅔ + 1⅔ + 0 + 6⅔ + 0	24

It should not be very surprising that the results for the average rankings are not very meaningful. We saw above that Person 1 and Person 2 agreed that parking was most important but disagreed on the relative importance of time and selection. Person 3 had a completely different ordering of attributes than did the other two subjects. For the investigator studying preferences in bookstores, the average results are not very useful. In opening a bookstore, for example, one would worry about which real individuals with their individual preferences would find the store attractive, not what some hypothetical average person would do. This example should make clear the problems of averaging preference scores. It is important to realize, however, that such averaging is common despite the fact that it is so ill advised.

FIGURE 3–10

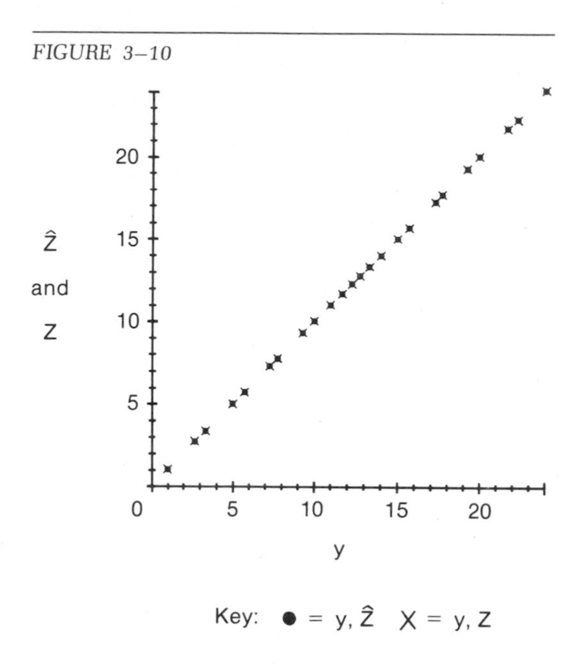

DATA REQUIREMENTS

The above simple examples suggest one important property of conjoint measurement. With only two possible levels for price, parking, and time and with three levels for selection, we had to ask respondents to rank 24 stores. The process is difficult. (It would be a useful exercise for the reader to make up 24 index cards, one for each combination, and then try to rank them.) With situations having more possible levels for the different attributes and perhaps having more attributes, the number of combinations grows quickly. To overcome the problems of large numbers of combinations (and the likely resultant fatigue or inattention of the respondent), investigators frequently use only a subset of the possible combinations (in what is called a fractional design). While very useful for reducing the size of the assessment task for input data, the use of fractional designs is dangerous, for the following reason. The preceding discussion emphasized that the loose restrictions involved in monotone conjoint measurement can cause serious problems in terms of how meaningful the results really are. Reducing the number of data points involved will exacerbate the potential problems. With fewer data points the fits will likely be very good but the actual meaningfulness of the fit and results becomes even more questionable.

A technique closely related to conjoint measurement is called *trade-off analysis*. In the examples above, we asked individuals to rate total packages of attributes (that is, we asked for ratings of bookstores, characterized on all four dimensions at once). In trade-off analysis, respondents consider two attributes at a time. For the bookstore example, they would be asked to give rankings from 1 to 4 in Table 3–19 comparing parking and time.

TABLE 3–19

	Parking	*No Parking*
5 minutes		
15 minutes		

They would similarly be asked to put numbers from 1 to 4 into the cells of a table for parking and price and into a table for time and price. They would give values from 1 to 6 to the cells of each of three tables involving selection. (See Table 3–20 for the array of tables.)

An algorithm very much like the conjoint algorithm described above would then be used to translate this input information into part-worths and Z values. Thus the major difference between trade-off analysis and conjoint measurement is in the form of the input judgments used. In practice, the choice depends in part on what computer program is available to a particular investigator and in part on which input form the investigator's subjects find easier to use.

SUMMARY

In summary, conjoint measurement is a technique for taking a dependent variable with an ordinal scale, together with independent variables for attributes, and fitting an additive model to explain the dependent variable in some ordinal sense. Of particular value is the fact that the resulting part-worths can be used to study the trade-offs among the independent variables. There are, however, two particularly important warnings that should be considered in using conjoint measurement. First, the basic model assumes additivity of effects. It does not, for example, consider balance among effects. Second, the loose restrictions of the conjoint algorithm pose potential serious problems, as discussed above.

TABLE 3–20

	Parking	No Parking
10% off		
List		

	5 Minutes	15 Minutes
10% off		
List		

	Very Wide	Wide	Medium
Parking			
No parking			

	Very Wide	Wide	Medium
5 minutes			
15 minutes			

	Very Wide	Wide	Medium
10% off			
List			

The first warning about additivity deserves a bit more attention. In considering some properties of some products it is reasonable to assume that additivity is a sensible approximation to reality. For other products or properties, however, the assumption is likely unsatisfactory. For example, consider automobiles. If we consider attributes of initial cost, luxuriousness, gas mileage, and similar measures, we may find additivity a reasonable assumption. If, instead, we focus on the car design and consider exterior color and interior color, additivity would likely be an entirely unsatisfactory assumption. For design, we care about balance. We cannot simply add the part-worths for a blue interior and a brown exterior to approximate the preference for the combination. The customer cares about balance and the conjoint model does not allow for the consideration of balance.

Finally, investigators should realize that conjoint measurements is similar to additive cross-tabs (main effects ANOVA). Both assume additivity. In conjoint, we require only an ordinal scale for the dependent variable. While users may find it attractive that conjoint applies looser restrictions on the dependent variable than does ANOVA, the looser restrictions are accompanied by serious potential problems in interpretability. While conjoint measurement is currently fashionable, investigators should use it with care. Sometimes they may want to run an ANOVA in addition to or in place of a conjoint analysis, on the theory that even if their dependent variables are not fully suited to ANOVA, at least the ANOVA approach avoids some of the problems of conjoint analysis.

CHAPTER 4

*Binary Regression**

Binary regression is a technique closely related to standard regression analysis, but intended for one type of situation in which the standard regression model is inappropriate. In particular, binary regression is useful when the dependent variable under analysis is a binary variable, taking only the values 0 and 1.

The ordinary regression model is

$$y = b_0 + b_1x_1 + b_2x_2 + \ldots + b_mx_m + e$$

where y is the dependent variable, $x_1, x_2, \ldots, x_m$ are independent variables, and e is a residual. In using this model, investigators are quite free to use independent variables of various types; for example, cardinal variables or dummy variables are both acceptable. Problems arise, however, when investigators want to use dependent variables which are restricted as to the range of values they can take.

LIMITATIONS OF ORDINARY REGRESSION

An example of such a situation might involve analysis of home-buying behavior. Suppose we are studying the question of whether or not families are likely to buy new homes in a one-year period within the same geographic areas in which they now live. (We are not considering long-distance moves.) Suppose that we collect data on a sample of families. For each we determine whether the family has moved in the past year or not. We might code the dependent variable 0 for those who have not moved and 1 for those who have. We could also collect information on potential explanatory or independent variables such as income, family size, and the like. We might want to fit an equation to our sample of families to explain their behavior. We could then use the equation to

*This chapter requires familiarity with the concept of likelihood, which is explained in Appendix A.

predict the home-buying behavior of additional families. The resultant predictions might be useful information for a real estate broker who wanted to identify attractive potential customers. In preparing these predictions, we would very much like to be able to interpret the predicted value of the dependent variable as a probability of moving. For that purpose, we would want to have predicted values in the range from 0 to 1. If we were to use a standard regression program to fit an equation to the data, however, we could not guarantee that the results would give predicted values in the desired range.

To see the problems with ordinary regression analysis for situations of this type, we now turn to another similar example. Suppose that we are trying to identify those married couples who are good prospects as Saturday evening customers for restaurants. Suppose that we have collected information on a set of couples, noting whether or not each couple went to any restaurant on a specific Saturday night. (Suppose we believe that that Saturday is a typical one.) This dependent variable is coded 1 for those who did go to a restaurant and 0 for those who did not. Suppose that for our first analysis we use total income for each couple as an explanatory variable. Table 4–1 provides a listing of a small set of observations. (As usual, we use fewer observations than we should in practice. The small database is used here for illustrative purposes.)

TABLE 4–1

Restaurant (1 if went to a restaurant; 0 if not)	*Income ($000)*
0	17
0	19
0	20
0	20.5
0	21.5
0	22
1	24
1	25
0	25
0	27
1	28
1	30
0	31.5
1	32
1	34
1	35
1	36.5
1	39
1	40
1	44

Figure 4–1 is a plot of the data from Table 4–1. Notice that the dependent variable takes on only the values 0 and 1, while the independent variable (income) takes on a range of values.

FIGURE 4–1

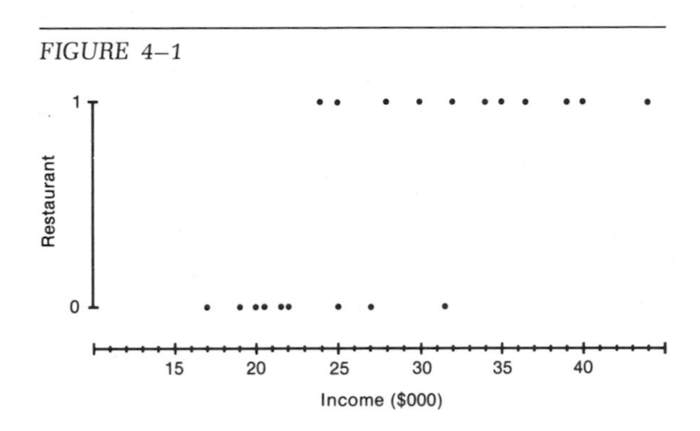

Just as it does with any other set of data, a regression analysis program would try to select a line for the configuration in Figure 4–1 that gave the best fit to the data. Figure 4–2 shows the results. Notice that the line slopes upward toward the right, as expected. The higher the income value, the higher the predicted value. If we interpret this predicted value as a probability of eating out on a Saturday night then, as expected, the probability increases with income.

FIGURE 4–2

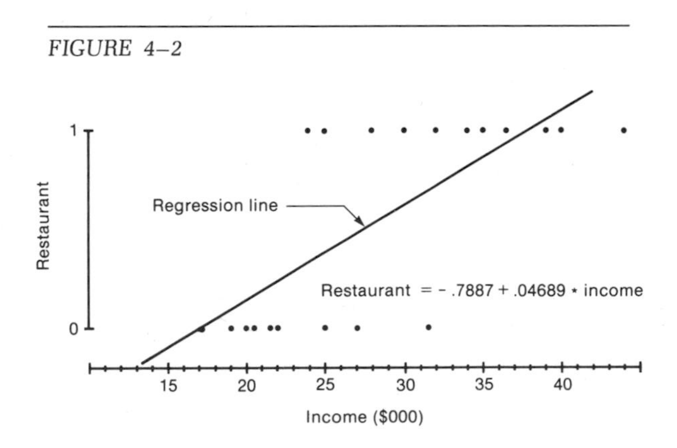

Now suppose that we use the line in Figure 4–2 to try to predict the behavior of additional couples. For a couple with total income of $25,000 the line indicates a probability of .38. For an income level of $41,000 it gives a probability of 1.13, however, which does not make sense. Similarly, an income of $14,000 corresponds to a probability of −.13, again out of the range of sensible values.

This example shows the most obvious problem of using standard regression for problems with binary dependent variables—the generation of predicted values outside the sensible range. In general, all regression equations may give unreasonable results when they are extrapolated to values for the independent variables far beyond the ranges of values to which the equations were fit. The problem with the line in Figure 4–2 is not of that sort, however. Notice that the line gives unreasonable predicted values close to and within the range of income values included in the set of input data. In using the equation to "predict" restaurant probabilities retrospectively for observations in the database, we would obtain predictors greater than 1 for three of those observations. Thus, the failure to produce predicted values in the range from 0 to 1 is a serious one when standard regression is used with binary data.[1]

BINARY REGRESSION

In principle the process of binary regression can perhaps best be thought of as simply adding one more step to the basic process of ordinary regression. (In fact, the calculations turn out to be quite different for the two techniques, but we defer consideration of the calculations in binary regression until later in this chapter.) For the current purpose, we can think of ordinary regression as follows:

Observations of the dependent variable are explained by the equation

$$y = \text{Predicted value} + e$$

where the prediction equation is of the form

$$\text{Predicted value} = b_0 + b_1 x_1 + \ldots + b_m x_m$$

and e is a residual. The regression procedure selects values for the coefficients in the prediction equation so as to achieve the best fit, in terms of smallest squared residuals, for observations in the database. It turns out that these coefficients are also maximum likelihood values under an assumption of normally distributed residuals.

[1]There are other problems with the use of standard regression for such situations. Such data violate the assumption of homoscedasticity in standard regression. With a binary dependent variable, the regression line will fit less well in the middle of the range of values taken in the database by the independent variable than it will toward the ends of that range. Yet, the assumption of homoscedasticity implies that the line fits equally well throughout that range. It turns out that in assuming homoscedasticity when in fact we know that the assumption is false, we are making less efficient use of the information in the database than we would by recognizing the heteroscedasticity that is present. Binary regression uses the information more efficiently.

The basic idea of binary regression is to add another step in finding the predicted values in an equation of the form:

$$y = \text{Predicted value} + e$$

For binary regression, the independent variables are used in a linear equation to find an intermediate quantity which we will call t. Unlike ordinary regression, however, binary regression does not assume that the prediction equation is such a linear form, because the values of t might lie outside the range from 0 to 1. Instead, in binary regression we use the t value to read another value, called $F(t)$, from a curve that by definition is restricted to the range from 0 to 1. Figure 4–3 gives an example of such a curve. Notice that it allows any value of t, positive or negative and large or small. The values of $F(t)$ all lie between 0 and 1.

FIGURE 4–3

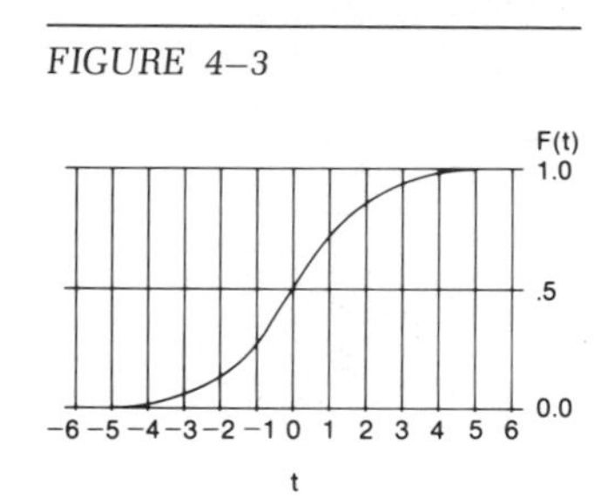

Thus, the basic idea of binary regression is to go from values of the independent variables to a value of t and then to go from the t value to a predicted value, $F(t)$, which is between 0 and 1:

$$t = b_0 + b_1x_1 + \dots + b_mx_m$$
$$\text{Predicted value} = \hat{y} = F(t)$$

As will be discussed further below, the fitting process in binary regression involves choice of the coefficients in the equation for t. In general, the user specifies which independent variables should be used. Users may also have some flexibility as to what curve to use for $F(t)$. With this type of procedure the predicted values can be interpreted as probabilities, restricted to the appropriate range.

To see in more detail how the procedure works, suppose that we have fit an equation

$$t = -9.632 + .3608 * \text{income}$$

to data on income and eating out. Suppose that we are using the $F(t)$ curve in Figure 4–3. Now, suppose that we want to estimate the probability of eating out for a couple with income \$25,000. We find a t value of $-9.632 + 25(.3608)$, or $-.61$. This t corresponds to an $F(t)$ of .35 as

FIGURE 4–4

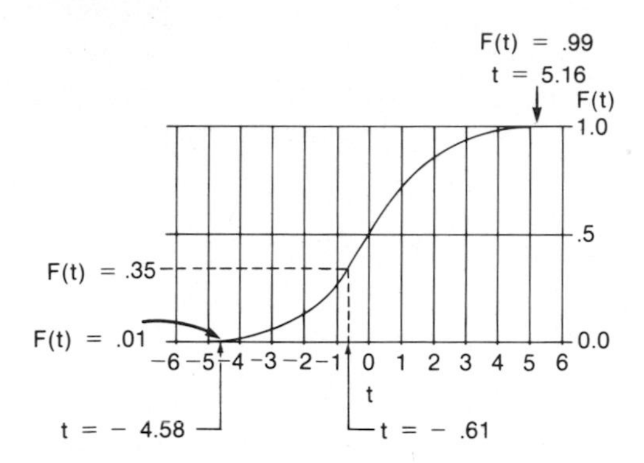

shown in Figure 4–4. Thus, we predict a .35 probability that the couple in question would eat out on the Saturday night under study. Similarly, incomes of $14,000 and $41,000 correspond to t values of −4.58 and 5.16, respectively. As shown in Figure 4–4, the corresponding probabilities are .01 and .99.

Binary regression can also be used with more than one independent variable. For example, consider the data in Table 4–2 on restaurant usage, income, and whether or not a couple must hire a babysitter in

TABLE 4–2

Restaurant (1 if went to a restaurant)	*Income ($000)*	*Babysitter (1 if need babysitter)*
0	17	0
0	19	1
0	20	0
0	20.5	0
0	21.5	0
0	22	1
1	24	0
1	25	0
0	25	1
0	27	0
1	28	0
1	30	0
0	31.5	0
1	32	1
1	34	0
1	35	0
1	36.5	0
1	39	1
1	40	0
1	44	1

order to go out to eat. (We assume that the last variable gives the couple's own judgment as to whether or not any children they have would require a babysitter.)

Binary regression analysis of the data from Table 4–2 with the curve in Figure 4–3 gives the following equation for t:

$$t = -9.456 + .3638 * \text{income} - 1.107 * \text{babysitter}$$

For a couple with income of $25,000 who would require a sitter, for example, we find a t value of -1.47 and a probability of .19. As is true in general in ordinary regression, we made an additivity assumption about the independent variables. In binary regression, however, we assume that the independent variables have additive effects on t, rather than on the predicted value $\hat{y}$ (as in ordinary regression). In this example, we assumed that the need for a babysitter did not affect t values for higher-income couples more or less than it did values for lower-income couples. Such an assumption is not likely accurate. If we thought that hiring a babysitter was a problem because of the expense, we could not justify this equation. If, on the other hand, we felt that for the people to whom we fit the model, and for the people whose behavior we might want to predict, the real constraint was availability of appropriate babysitters on Saturday nights rather than expense, we might find the additivity assumption an acceptable approximation of reality.[2]

INTERPRETATION OF OUTPUT

The binary regression model can be extended to consider other independent variables. The model has the considerable advantage that it gives predictions that can be interpreted as probabilities. However, the restriction of predicted values to the range from 0 to 1 creates some unavoidable disadvantages in interpretation of the coefficients. If we had fit an ordinary regression equation to the data in Table 4–2, we would interpret the coefficient of income as the change in the dependent variable per unit change in income. Similarly, the coefficient for babysitter would be the change in the dependent variable caused by the need for a babysitter. In ordinary regression, the additivity assumption means that the coefficient for income is the change in $\hat{y}$ for each unit change in income—for the change from $20,000 to $21,000 as well as for the change from $30,000 to $31,000 or the change from $39,000 to $40,000.

With binary regression, the linear equation is used to give values of t, not values of $\hat{y}$. Thus, the coefficients are the changes in t per unit

---

[2]As in ordinary regression, the investigator can include specific interaction terms (generally as products of variables) in the linear form in binary regression.

changes of the corresponding independent variables. In the equation for the data from Table 4–2, the value of t changes .3638 for each unit change in income. The problem is that we are really not interested in t. We want to know about the effects of income on $\hat{y}$. (We are using binary regression simply to ensure that the predicted values for y lie in the appropriate range.) Even though each unit change in income gives the same change in t, each such change in t does not give the same change in $\hat{y}$. (Notice that the curve for $F(t)$ rises considerably more steeply in some ranges of t values than it does in other ranges.) Thus, interpretation of coefficients in binary regression must be more complex.

We can, of course, examine the signs of the coefficients to see if they make sense. An increase in t will always correspond to an increase in the predicted value $\hat{y}$, so a positive coefficient means that $\hat{y}$ increases as the independent variable increases, while a negative coefficient means $\hat{y}$ decreases when the independent variable increases.

In order to consider information on how much $\hat{y}$ changes when we change an independent variable such as income, however, we have to specify a starting value for $\hat{y}$. The reason is that $\hat{y}$ (which equals $F(t)$) changes with t (and thus with the terms that make up t) at different rates in different ranges of values of t. For example, in the equation based on Table 4–2, suppose that we start from $\hat{y} = .726$ (which is equivalent to $t = .6$) and that we increase income by $1,000. t will increase by .3638 (to .9638) and $\hat{y}$ will increase to .832. The difference in $\hat{y}$ with this particular $1,000 change is .106. If instead we had started from $\hat{y} = .846$ (or $t = 1.020$) and increased income by $1,000, t would still increase by .3638. In this case, however, the new t (1.3838) corresponds to a $\hat{y}$ of .917. The difference in $\hat{y}$ with this $1,000 change is .071. Thus, we can talk meaningfully about changes in $\hat{y}$ with changes in the independent variables only if we specify a starting point. For example, we might be able to obtain output from a program for binary regression on the change in $\hat{y}$ (predicted probability) per unit change in income starting from a $\hat{y}$ value of .5.

COMPUTATIONS FOR BINARY REGRESSION

We can now turn to the question of how a computer program performs the binary regression analysis. The actual fitting procedure involves the concept of likelihood. It also involves a search to find the best set of coefficients, in terms of likelihood. Suppose for example that, for the data in Table 4–2 and the $F(t)$ curve in Figure 4–3, a program is investigating the values -6 for b_0, $+.3$ for b_1, and -1.2 for b_2. Thus it is considering the possible equation

$$y = F(t) + e$$

$$t = -6 + .3 * \text{income} - 1.2 * \text{babysitter}$$

This possible equation can be used to find values of t for each observation in the database, as shown in Table 4–3. The t values can then be converted to $F(t)$ values by reading Figure 4–3. Table 4–3 summarizes all of these calculations. For example, the first t value is found from

$$t = -6 + .3 * 17 - 1.2 * 0 = -.9$$

This t is seen in Figure 4–3 to correspond to an $F(t)$ of .29 and the corresponding $1 - F(t)$ is .71.

Now, if the possible equation under consideration is in fact a correct one, then the last two columns of Table 4–3 gives the probabilities of y values of 1 and 0, respectively. Thus, on a particular observation the probability that y is 1 conditional on the values -6, .3, and -1.2, for b_0, b_1, and b_2 is:

$$pr(y = 1 \mid -6, .3, -1.2) = F(t = -6 + .3x_1 - 1.2x_2)$$

and the conditional probability that y is 0 is

$$pr(y = 0 \mid -6, .3, -1.2) = 1 - F(t = -6 + .3x_1 - 1.2x_2)$$

TABLE 4–3

Observation	*Restaurant*	*Income*	*Babysitter*	*t*	*F(t)*	*1 − F(t)*
1	0	17	0	−.90	.29	.71
2	0	19	1	−1.50	.18	.82
3	0	20	0	.00	.50	.50
4	0	20.5	0	+.15	.54	.46
5	0	21.5	0	+.45	.61	.39
6	0	22	1	−.60	.35	.65
7	1	24	0	+1.20	.77	.23
8	1	25	0	1.50	.82	.18
9	0	25	1	.30	.57	.43
10	0	27	0	2.10	.89	.11
11	1	28	0	2.40	.92	.08
12	1	30	0	3.00	.95	.05
13	0	31.5	0	3.45	.97	.03
14	1	32	1	2.40	.92	.08
15	1	34	0	4.20	.99	.01
16	1	35	0	4.50	.99	.01
17	1	36.5	0	4.95	.99	.01
18	1	39	1	4.50	.99	.01
19	1	40	0	6.00	1.00	.00
20	1	44	1	6.00	1.00	.00

Suppose that y is actually 1 on an observation. Then the likelihood of the current set of coefficients for that one observation is

$$pr(y = 1 \mid -6, .3, -1.2)$$

Similarly, the likelihood for an observation on which y is actually 0 is

$$pr(y = 0 \mid -6, .3, -1.2)$$

To find the likelihood for the entire sample, we simply multiply together the likelihood values for the individual observations. Table 4–4 shows the procedure. For each observation $F(t)$ has been circled if y is 1 and $1 - F(t)$ has been circled if y is 0. The likelihood is the product of the circled values, or .0000235:

$$.71 * .82 * .50 * .46 * .39 * .65 * .77 * .82 * .43 * .11 * .92$$
$$* .95 * .03 * .92 * .99 * .99 * .99 * .99 * 1.00 * 1.00$$

Other sets of coefficients would give other likelihood values. The basic procedure in binary regression is to search for the best set of

TABLE 4–4

Observation	Restaurant	F(t)	1 − F(t)
1	0	.29	(.71)
2	0	.18	(.82)
3	0	.50	(.50)
4	0	.54	(.46)
5	0	.61	(.39)
6	0	.35	(.65)
7	1	(.77)	.23
8	1	(.82)	.18
9	0	.57	(.43)
10	0	.89	(.11)
11	1	(.92)	.08
12	1	(.95)	.05
13	0	.97	(.03)
14	1	(.92)	.08
15	1	(.99)	.01
16	1	(.99)	.01
17	1	(.99)	.01
18	1	(.99)	.01
19	1	(1.00)	.00
20	1	(1.00)	.00

values for the coefficients, where "best" means that set giving the highest likelihood value. At each step, the procedure would evaluate the current set of values, as shown in Tables 4–3 and 4–4. It would calculate the likelihood value and would determine whether changes in the coefficients would improve the likelihood appreciably. If so, it would change the coefficients and proceed to evaluate the new set.

It is useful to be aware of one potential problem in procedures for binary regression. Consider the data in Figure 4–5, showing y plotted against x for a set of data.

FIGURE 4–5

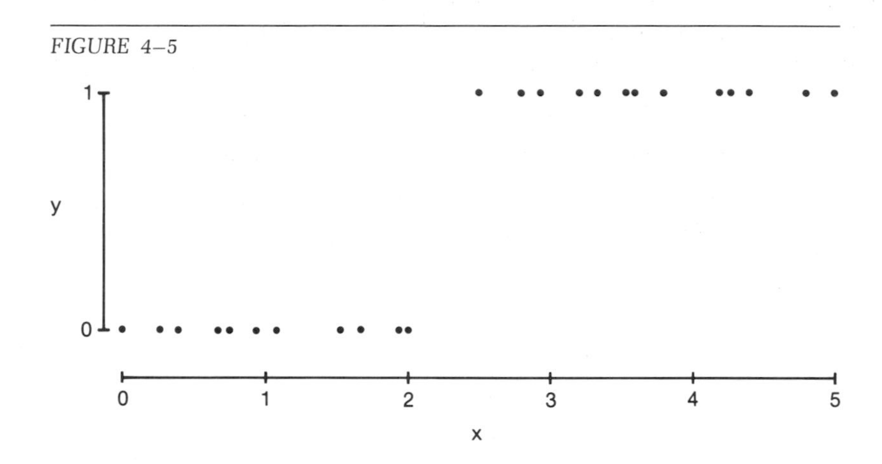

Notice that y is 0 for all xs of 2 or below and that y is 1 for all xs of 2.5 or more. The maximum likelihood value of b_1 will be infinite for these data (giving $\hat{y}$s of 0 below some dividing value between 2.0 and 2.5 and $\hat{y}$s of 1 above that value). Analogous situations can arise in more than two dimensions. Some binary regression programs will report an inability to solve such problems. Other programs will simply search for a long time for the maximum likelihood value(s) and may output one or more large values. (In general, users of specific binary regression programs should check to see how the programs handle such situations.)

To specify the binary regression procedure, it remains to specify the form of the curve for $F(t)$. We need a form that will work for any value of t and produce values of $F(t)$ between 0 and 1. Cumulative probability distribution curves are well suited for this purpose. One common choice is the unit normal distribution (the normal probability distribution with mean 0 and variance 1). When such a curve is used, the procedure is called *probit analysis*. An alternative is to use the logistic cumulative distribution curve, given by the equation

$$F(t) = \frac{1}{1 + e^{-t}}$$

When this curve is used, the procedure is called *logit analysis*. While there are theoretical differences between the two, in practice the two forms give analogous results and the choice between them is not considered important.[3]

On computational grounds, the logit procedure is faster, because it is computationally easy to translate between t and $F(t)$ for the logistic curve. In addition, with the logit model the coefficients have a somewhat clearer interpretation than they do with other forms of binary regression. In logit analysis, the linear function of the independent variables gives the log odds in favor of a y value of 1. In other words

$$\log \frac{pr(y=1)}{pr(y=0)} = b_0 + b_1 x_1 + \ldots$$

Thus, the coefficients are the changes in the log odds per unit change in the corresponding independent variables. While not as easily interpretable as are the coefficients in ordinary regression, these logit coefficients do sometimes allow useful interpretability.

The output from a binary regression includes the coefficients found for the independent variables. Programs may also provide summary information on the likelihood level attained or on R^2, which is entirely analogous to the value from ordinary regression.[4] Table 4–5 shows the calculation of R^2 for the data from Table 4–2.

Binary regression programs may also provide information summa-

---

[3]Probit analysis was devised for cases of the following sort: Suppose that a group of experimental animals have unobservable tolerance or resistivity levels for some drug. Suppose animal i has resistivity r_i. Suppose that the r levels are normally distributed in the population of animals. Suppose further that the dosage or toxicity level t received by an animal is not observable, but that t is a linear function of certain observable variables $x_1, x_2, \ldots$. For example, the xs might be measures of the amounts ingested of various substances.

Thus

$$t = b_0 + b_1 x_1 + b_2 x_2 + \ldots$$

Suppose that an animal dies if its dosage is higher than its resistivity level. If t_i is the dosage for animal i, then the animal dies only if $t_i > r_i$.

The essence of this example is that the binary dependent variable y takes the value 1 only if t_i exceeds the value of r_i. r_i is normally distributed. t_i is a linear function of the independent variables (the xs). For such examples, there is theoretical justification for use of the probit model, although in practice the logit can also be used successfully. (In actuality, the probit model fits a somewhat wider class of problems than this statement suggests. It turns out that the mean of r can also be a linear function of the xs. In addition, the values of t given above may be means of normal distributions for t values rather than constants as specified here.)

Thus, probit and logit analysis do have some different characteristics. As noted above, however, the two types of analysis give similar results. Since logit analysis costs a bit less, it is often preferred to probit analysis.

[4]Programs may provide additional information about the search for the maximum likelihood values.

TABLE 4–5 *

Observation	y = restaurant	F(t)* = "predicted" probability	y − predicted value = e (residual)	e^2
1	0	.037	−.037	.001369
2	0	.025	−.025	.006250
3	0	.102	−.102	.010404
4	0	.119	−.119	.014161
5	0	.163	−.163	.026569
6	0	.072	−.072	.005184
7	1	.326	.674	.454276
8	1	.411	.589	.346921
9	0	.187	−.187	.034969
10	0	.591	−.591	.349281
11	1	.675	.325	.105625
12	1	.811	.189	.035721
13	0	.881	−.881	.776161
14	1	.746	.254	.064516
15	1	.949	.051	.002601
16	1	.964	.036	.001296
17	1	.979	.021	.000441
18	1	.974	.026	.000676
19	1	.994	.006	.000036
20	1	.996	.004	.000016
	V(y) = .248			V(e) = .112

$$R^2 = 1 - \frac{.112}{.248} = .55$$

*t is −9.456 + .3638 * income − 1.107 * babysitter.

rizing a process in which each observation in the database is classified retrospectively (with the regression equation) as a 0 or a 1. For example, a program might classify as 1s those observations with $F(t)$ values greater than .5. It could then provide summary output comparing actual values of the dependent variable with these classifications, telling how many observations were correctly and how many were incorrectly classified as 0s and 1s.

CHAPTER 5

Discriminant Analysis*

Discriminant analysis is a multivariate technique whose end purpose generally is to provide a procedure for classifying individual observations into one of a set of groups or populations. The discriminant procedure involves starting with a set of observations whose group memberships are known. That initial set of data is used to fit (or calibrate) a relationship which can subsequently be used to classify other observations whose group memberships are not known. Thus, investigators must begin the procedure knowing what groups are relevant for their analysis. They must also begin with (random) samples of observations from each of the relevant groups. For each observation, investigators must have information on group membership and also on the values of one or more *discriminator variables* which can be used to establish a classification rule. The discriminant procedure fits such a rule. For any subsequent observation, the investigators must know values for the discriminator variables only. They then use the results of the discriminant analysis to determine the profitabilities of the new observation's membership in each of the groups.

The basic purpose of discriminant analysis can be explained further by examples of situations in which the technique might be appropriate. Kendall[1] gives several reasons why discriminant procedures might be used. In some cases investigators want to predict future group membership. An example is in credit rating. Before granting credit to an individual, banks, finance companies, and other organizations would like some assurance that the individual is likely to be a good credit risk. One procedure for identifying potential good and bad risks is to select some discriminator variables such as income, amount of

*This chapter uses the concept of likelihood, which is explained in Appendix A. That appendix also provides a brief review of Bayes' Theorem.

[1]Maurice Kendall, *Multivariate Analysis* (New York: Hafner Press, 1975).

previous credit, length of time employed, and so on. Samples of people whose credit behavior is known are used to calibrate a relationship for using values of these discriminators to predict credit worthiness. The relationship can then be used as part of a credit screening for new applicants whose values on the discriminator variables are known but whose future credit behavior obviously is not yet known.

Discriminant analysis can also be appropriate in industrial or laboratory situations in which testing involves destruction of the object tested. Investigators would select as potential discriminator variables measurements that could be taken without destroying the object. They would then provide information about a sample of observations on which actual test results were known. (Those objects would be subjected to the destructive test to determine this information.) If discriminant analysis could use that sample to establish a relationship for predicting the test results from the values of the discriminator variables rather than from the actual test, then additional objects not subjected to the test could be analyzed and the procedure could predict how those objects would be likely to perform if they were in fact tested fully.

Kendall also mentions applications to situations involving lost or unobtainable information. For example, in archeology investigators must often try to identify the gender of a partial skeleton that is discovered. Using information on other findings of more complete skeletons, they can establish discriminant procedures for classifying partial remains as male or female. Situations involving unobtainable information are encountered in medical applications. Clinicians can attempt to identify relatively easily obtained measures that are reflections of other useful measures which are harder to obtain. For example, they may use external measurements and the results of blood tests to try to identify conditions which otherwise could only be found through risky surgery.

A related purpose of discriminant analysis is illustrated by the example used later in this chapter. That example involves group membership which could be found but only at some expense. As a result, a discriminant procedure is suggested as a screen for identifying likely prospects. More detailed and costly examination can then be applied to those prospects. The example used below involves the distribution method chosen for various industrial products. We assume that a group of marketing consultants specializes in working with manufacturers to establish effective direct distribution for industrial products. In order to find new business, the consultants would like to find products which are good candidates for distribution through direct channels, even though the products are currently being distributed through indirect ones (distributors). Such products would be likely prospects for the consultants' services if the manufacturers were interested in establishing direct distribution. The consultants could spend additional sales time with such likely prospects to determine whether direct distribution would in fact make sense for the products; if so, the consulting services might be valuable to the manufacturers. Notice that this example is a slight variation on the usual group-membership prediction. A discriminant relationship would be fit with observations whose group (distribution methods) were considered appropriate. The relationship would then be used to predict the distribution methods of additional products whose methods were known, but not known to be appropriate. A mismatch between the predicted and actual methods for one of these products would identify the product's management as a possible prospect for the consultants.

ONE DISCRIMINATOR

Before further consideration of this example, it is useful to consider some simpler illustrations of the basic ideas of discriminant analysis. First, a bit of terminology. In a database for discriminant analysis, the variable giving the group membership of each observation is called the *identifier variable*. That variable will generally be a dummy or categorical variable. In one problem, the variable might be coded 1 for female and 0 for male. In another problem, the identifier might be coded 0 for employed full time, 1 for employed part time, and 2 for self-employed. As mentioned above, the *discriminator variables* are used to establish the relationship for predicting group membership. There can be one or more than one discriminator variables. It is customary in discriminant analysis to call the identifier variable Y and the discriminator variables X_1, X_2, and so on.

To begin, we consider the case of a single discriminator X. Suppose that we have data on a group of MBA students. For each student we know gender (0 for male and 1 for female) and height (X). Suppose that for some reason we want to predict gender from height for additional MBA students. The groups or populations are male MBA students and female MBA students.

A basic assumption of discriminant analysis is that the discriminator variable(s) follow a known type of probability distribution in each group. In the case of a single discriminator X, the assumption is that the X values in each group follow normal distributions. In the current example, we would assume that the heights of male MBA students followed a normal distribution and that the heights of female MBA students followed another normal distribution. The familiar bell-shaped normal distribution is characterized by a mean and a standard deviation (or equivalently, by a mean and a variance). For reasons that are not intuitively obvious, the most common assumption in discriminant analysis is that the normal distribution for each of the groups or populations has the same variance. With more than one discriminator variable the basic assumption of analysis is that the discriminators follow multivariate normal distributions in each group and the usual assumption is that the variance structures (the variances and covariances) are the same in each group. For the time being, we will pursue the example on MBA students under the assumption of equal variances for males and females. Later, the chapter will explain further why such an assumption is the standard one.

There are two basic tasks in applying discriminant analysis in this example. First, the available data are used to estimate the normal distributions for males and for females. Then, the distributions so identified are used for purposes of reporting results and classifying new individuals. For this chapter it is appropriate to emphasize the second of these tasks. In fact, the fitting of the distributions is a mechanical procedure performed by a computer program, and for the user of discriminant analysis the main points to understand concern the use of the resulting distributions.

Users of discriminant analysis must, however, be satisfied that their data meet the basic requirements for the fitting procedure to work. The data must include a random sample from each membership group. Those group samples will then be considered representative (of their groups) in the fitting procedures. This requirement of discriminant analysis is not present with many other techniques for multivariate data analysis. For example, in regression (and binary regression), investigators often purposely select values of the independent variables to cover interesting and useful situations; they do not need to provide the procedures with representative samples with regard to those variables.

Once users are convinced that their data are appropriate, they need not worry about the details of the fitting. Thus, for the time being, assume that we know the true distribution of the heights of male MBA students and the true distribution of the heights of female MBA students. Assume that those distributions are as shown in Figure 5–1. The

FIGURE 5–1

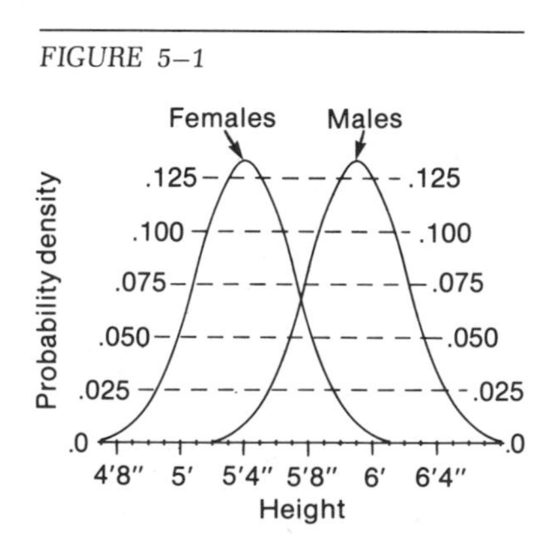

figure depicts the density function for each distribution. Recall that for such a function the height of the curve above any range of values is proportional to the probability of that range of values. The curves show that the mean height of male MBA students is 5 feet, 11 inches, while that of female MBA students is 5 feet, 4 inches. Both distributions have standard deviations of three inches (variances of nine inches squared).

The density functions in Figure 5–1 are likelihood functions for the current problem. In other words, for information on a person's height (which can be considered a piece of sample information), the graphs give the conditional probability densities of that height, given that the student is male (Pr (height | M)) and of that height given that the student is female (Pr (height | F)). Recall that these conditional probabilities are defined as the likelihoods of male and female, respectively. For example, for a height of 5 feet, 11 inches the likelihood of male is .1330 while the likelihood of female is .0087. For a height of 5 feet, 9 inches the likelihood of male is .1067 while the likelihood of female is .0332. For a height of 5 feet, 7 inches the likelihood of male is .0547 and the likelihood of female is .0807. (Figure 5–1 clearly cannot be read with this much precision. These values were obtained from a table of the normal density function.)

CLASSIFYING ADDITIONAL OBSERVATIONS

Suppose that we know that Figure 5–1 gives the true likelihood information and that we want to classify an individual as male or female. In fact, we would really like to determine the probability that the person is female and the probability that s/he is male. Suppose s/he is 5 feet, 7 inches tall. One simple way to perform such a classification is just to select that group with the higher likelihood value for the 5 foot, 7 inch–height. With this rule, we would classify the person as female. We might further report the odds of female versus male, using the likelihood values. The odds ratio would be 807/547.[2]

This procedure is equivalent to dividing the range of heights at the point of equal likelihood (5 feet, 7½ inches). For heights above this point, students would be classified as male. For heights below 5 feet, 7½ inches they would be classified as female. It is useful to ask what part of each probability distribution is on its respective side of this dividing point. Tables of the normal distribution tell us that only .122 of the distribution for male heights lies to the left of 5 feet, 7½ inches (which is 1⅙ standard deviations below the mean). Similarly, only .122

[2]*Odds* thus simply give a ratio of probabilities. Often, to avoid decimal points, both numerator and denominator in such a ratio are multiplied by the same power of 10. Without such a multiplication the ratio in the text would be written as .0807/.0547.

FIGURE 5–2

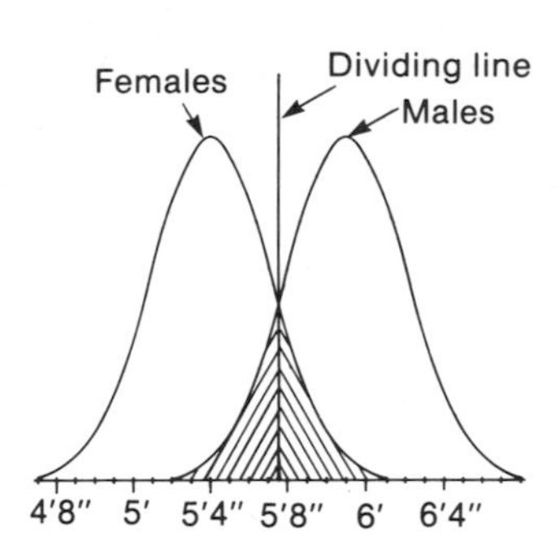

of the distribution for female heights lies to the right of that value. Thus, if we classify individuals according to this simple rule, we can expect to classify correctly .878 of the female MBA students and .878 of the males (.878 is 1.0 − .122). Figure 5–2 shows the portions of each of the two probability distributions for which we would make mistakes in classification with this rule. (Those portions are shaded in the figure.) The .878 figure tells what fraction of each distribution is on its own side of the dividing line in that figure. The value is called the *separation* of the discrimination.

This suggested procedure for classification is certainly simple to apply but it has drawbacks. One obvious problem is that the procedure does not consider the relative fractions of males and females in the MBA student body. In the early 1970s at Harvard Business School, only about 5 percent of the students were female. Even if the distributions in Figure 5–1 were correct for male and female students, the classification procedure given above would be unsatisfactory if the people to be classified had been students at that time. The 5 foot, 7 inch–person described above might well have been a male. While the height is a bit more characteristic of females, there were many more male students and the height was not particularly uncommon for males.

Thus, the simple classification procedure can be improved by considering additional information on the relative probabilities of each of the two groups. In the terminology of statistics, such values are called *prior probabilities*. The suggestion here is to use both prior information and likelihood information to determine classification probabilities (or *posterior* information). It should seem reasonable to the reader that the procedure involved is an application of Bayes' theorem. For example, suppose that we are classifying MBAs from the class of 1979, in which approximately 20 percent were female. The prior probabilities are then .2 for female and .8 for male. Figures 5–3 and 5–4 below show how

FIGURE 5–3

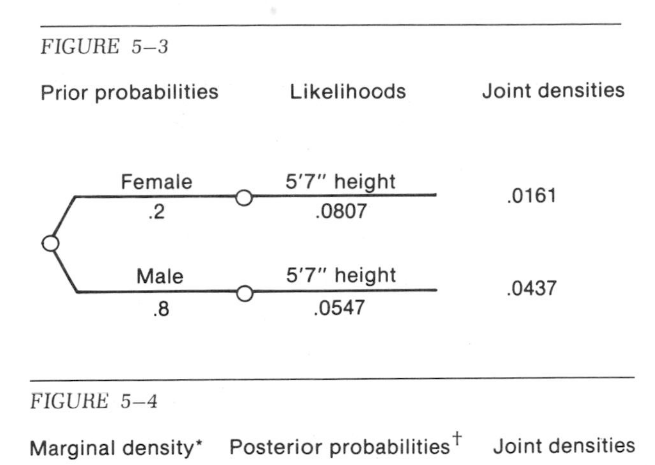

FIGURE 5–4

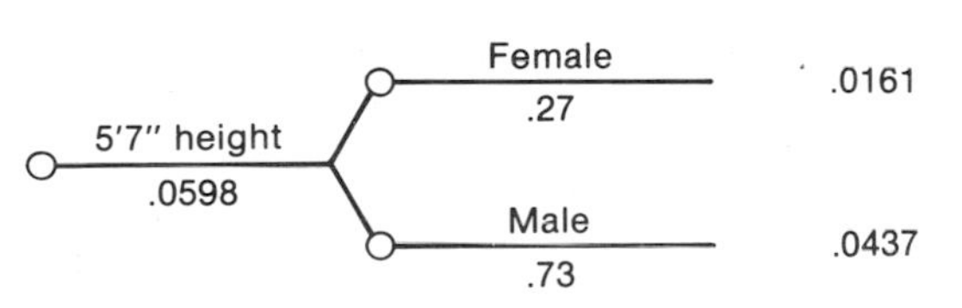

* The marginal density is found by adding the two joint densities.

† The posterior probabilities are found by dividing the marginal density into the joint densities.

Bayes' theorem is used to find the posterior probabilities of group membership for a member of that class who is 5 feet, 7 inches tall. Even though the height is a bit more characteristic of females than it is of males, the predominance of males in the class makes the posterior probability of male considerably higher (.73 versus .27). The odds in favor of female are only 27/73.

It is generally possible in discriminant analysis to apply such prior probabilities of group membership in classification. Even statisticians who are non-Bayesians apply Bayes' theorem in such cases. The first procedure considered above, in which classification was performed on the basis of likelihood values only, is appropriate only for cases with equal prior probabilities. In such cases the posterior probabilities are proportional to the likelihoods.

To be really correct, we should add one further dimension to the analysis. In real-world discriminant analysis problems, there are often different costs of making different types of classification errors. The point is best illustrated with an example such as credit scoring, where the cost of rejecting someone who would have been a good risk is not the same as the cost of accepting someone who proves to be a bad risk.

TABLE 5–1

	Classified as: M	Classified as: F
True Gender: M	0	C_{FM}
F	C_{MF}	0

In the first case, the cost may be the lost interest on the loan (unless there are many other good risks available, in which case there may be little, if any, cost). In the second case the cost may be a substantial portion of lost principal of the loan.

Thus, the classification procedure involving both prior and likelihood information can often be improved by adding cost information as well. Suppose for the MBA-student example that the costs are as shown in Table 5–1. There is no cost for a correct classification. C_{FM} is the cost of classifying someone as female whan he is in fact male. C_{MF} is the cost of classifying someone as male when she is in fact female. For illustrative purposes, Figure 5–5 shows how the costs can be incorporated with the probability information in an analysis to find the expected cost of classifying a 5 foot, 7 inch–tall individual as male and the ex-

FIGURE 5–5

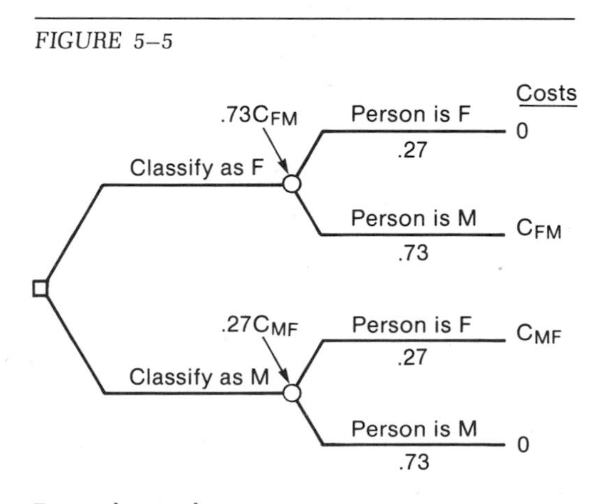

Expected costs of:
Classifying as F = $.73 * C_{FM}$
Classifying as M = $.27 * C_{MF}$

pected cost of saying that the same person is female.[3] To make the classification with lowest expected cost, we would classify the person as F if $.73 * C_{FM}$ were less than $.27 * C_{MF}$. Otherwise we would classify the person as M. (This example should be taken primarily as an illustration of the mechanics involved. Since the purpose of this classification has not been explained, the reason for using costs and the nature of the costs are not as clear as would be true for the credit scoring example.)

Most programs for discriminant analysis do allow use of both likelihood and prior information in classification. Most programs do not, however, consider costs of misclassification. Users can apply the type of analysis illustrated in Figure 5–5 to incorporate costs.

MEASURES OF FIT

Before leaving the case of a single discriminator, we return to the usual case in which the distribution for each group is not known for certain. In such cases the discriminant program must take data on observations whose group memberships are known and use those data to estimate each distribution. Earlier, this discussion argued that the procedure is a relatively straightforward technical one and the user of discriminant analysis need not know the details. It is appropriate, however, to consider one additional measure of fit of the discrimination that is based on the input data. The first measure, called the separation, was described above. It tells what fraction of each probability distribution lies on its own side of the dividing line. Ths second set of information on fit is called a *confusion matrix*. The discriminant program uses the relationship it has fit to classify each observation in the database retrospectively. In general, the prior probabilities used for this purpose are the

---

[3]Figure 5–5 contains a *decision tree* (as opposed to a *probability tree* of the sort used in Figure 5–3 and described in the appendix material on Bayes' Theorem). The small square at the left of Figure 5–5 is a *decision node;* the decision maker in the problem must choose one of the branches leading from that node (in other words, must classify the individual as F or else classify the individual as M). The small circles at the ends of the decision branches are called *event nodes;* exactly one of the branches coming from an event node will turn out to be true, but the decision maker cannot select which outcome it will be. (The person will be F or will be M.) The branches leading from the event nodes are labeled with their respective probabilities (.27 and .73). The *endpoints* of the tree are labeled (in the column headed "costs") with the costs of following the paths to those endpoints. (For example, the second endpoint from the top is the cost of classifying someone as F who is really M.) The event nodes are labeled with the probability-weighted (or *expected*) *costs* of being at those nodes. (For example, for the top event node the expected cost is $.27(0) + .73(C_{FM})$, or $.73 * C_{FM}$.) The expected costs are also listed below the tree.

For a considerably more detailed discussion of decision tree analysis, the reader might consult *Quantitative Methods in Management* by Vatter, Bradley, Frey, and Jackson (Homewood, Ill.: Richard D. Irwin, 1978).

fractions of the database in each group. (Because the input data must contain the group membership for each observation, these fractions are easily found.) The confusion matrix lists true group on one side and classified group along the other. It lists the number of observations in each cell. For example, Table 5–2 is a hypothetical confusion matrix for the MBA example. The table says that 75 males were correctly classified while 5 males were incorrectly classified as female. Similarly, 12 and 8 female MBA students were correctly and incorrectly classified, respectively.

TABLE 5–2
Confusion Matrix

		Classified as:		
		M	*F*	*Total*
True Group	*M*	75	5	80
	F	8	12	20
	Total	83	17	100

In using confusion matrices it is important to realize that the fractions classified correctly and incorrectly do not necessarily reflect what will happen when additional observations are considered. For one thing, the database may not reflect the same proportions as the population from which new observations will be drawn. In fact, if one of the groups is small, it may be good practice to include a disproportionate number of members of that group in the database to calibrate the relationship well. If so, the database would certainly not be representative of the population.[4] A second reason why the confusion matrix cannot predict future misclassification problems exactly is always present. The matrix shows the classification results for the same observations on which the discriminant analysis was performed. For that reason, the results in the matrix will in general be better than would be the results for a set of observations not used in the fitting.

[4]It is important to distinguish this statement from the one on page 92 stating that discriminant analysis assumes representative samples *within* each group. The current statement says that the numbers of observations from various groups need not be in proportion to the sizes of those groups in the population. Thus, there need not be representativeness *across* group samples—but there must be representativeness *within* each such sample.

EQUAL VERSUS UNEQUAL VARIANCES

Before leaving the topic of discriminant analysis with one discriminator variable, we return now to the issue of equal versus unequal variances. On first encountering discussions of discriminant analysis, essentially everyone finds the common requirement of equal variances surprising and restrictive. It seems more sensible at first to allow the groups to have unequal variances. The reason for the usual assumption of equal variances is illustrated in Figure 5–6, where the density functions for two hypothetical groups with unequal variances are shown. Discriminant analysis in one dimension usually finds a dividing value between groups. Notice that such a procedure is not possible in Figure 5–6, where both high and low values are characteristic of Group B, while intermediate values are more characteristic of Group A. A rule for classification even on the basis of likelihoods alone in this case cannot take the simple form: Put all observations with values below the dividing value in one group and all observations with values above the dividing value in the other. The mathematics involved in fitting sensible results for such a case are considerably complicated over the simpler case of equal variances. With more than one discriminator, the situation is analogous when users do not assume equal variance structures (variances and covariances among the discriminators). For this reason, most computer programs for discriminant analysis do not even allow the user the option of assuming unequal variance structures. (A few programs do allow the option.)

FIGURE 5–6

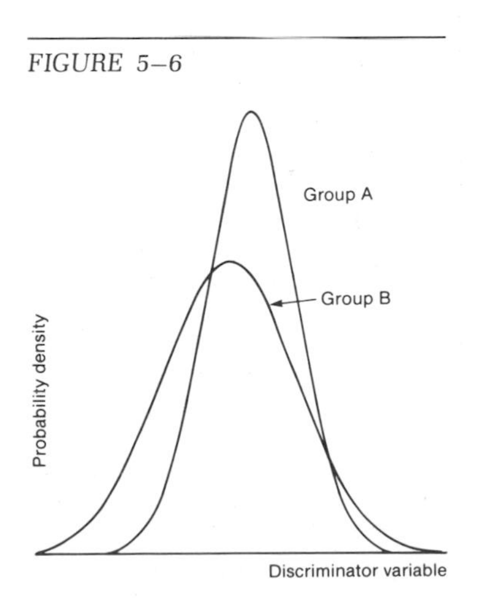

DISCRIMINANT ANALYSIS WITH TWO DISCRIMINATORS

With two discriminator variables, the basic process of discriminant analysis remains much the same as the one described above for a single discriminator. The user presents a computer program with (representative) information on observations whose group memberships are known. For the time being we will consider only two groups. The user must also give values of the discriminator variables. The program estimates the multivariate normal distribution of the discriminator variables in each group. Again, the fitting procedure is a technical but well-defined one and the user of discriminant analysis need not know the details. After fitting the distributions, programs generally report results in several ways. They give information about the discriminant function, a function of the discriminator variables that separates the two groups as well as possible. This function is analogous to the dividing line in Figure 5–2. The programs generally also provide information on the extent of separation of the two groups, and they provide a confusion matrix. These outputs are illustrated below for the problem on direct and indirect distribution described earlier in the chapter. First, however, graphical descriptions of discriminant analysis with two discriminators are included for those readers who find such descriptions useful.

Figure 5–7 shows a plot of a hypothetical set of input data, with a

FIGURE 5–7

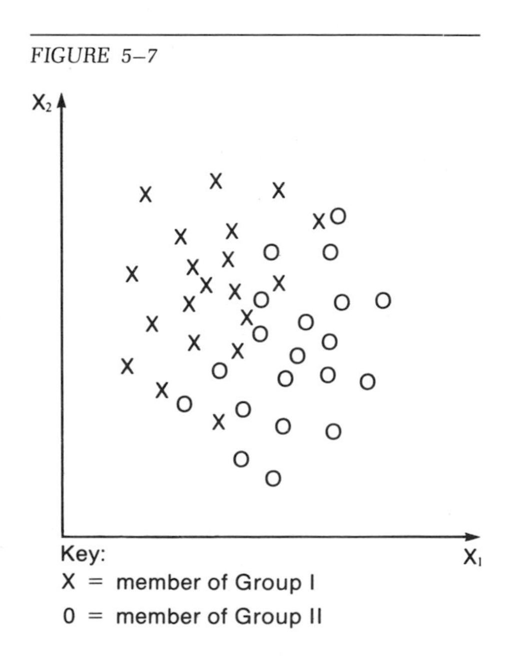

cross for each observation in Group I and a circle for each observation in Group II. The points are plotted according to their values on the two discriminator variables X_1 and X_2.

Discriminant analysis programs fit multivariate normal distributions to such sets of input data. With the assumption of equal variance structures, the distributions for the two groups must have the same shape, although they will be located at different points in the plane of X_1 and X_2. For the multivariate normal distribution the set of points with the same density (or likelihood) value all fall along a closed ellipsoid-shaped curve. Two such curves are shown in Figure 5–8. That figure gives the results of fitting distributions to the data in Figure 5–7. One of the ellipsoids gives all points with a specific density value for the distribution for Group I. The other closed curve gives the points for the distribution from Group II with the same density value. Notice that the ellipsoids have the same shape but that they are located slightly differently in the plane of the figure.

With two discriminators, the idea in separating the groups is to find a line in the plane to separate them. This step is analogous to that of finding the dividing value in Figure 5–2. Figure 5–9 shows such a dividing line, shown there as a dashed line. The basic idea of classification will now involve this line. On the basis of likelihood values only, the procedure would be to classify all the observations which fall to the upper left of the line as belonging to Group I and all the observations to the lower right of the line as in Group II. (As in the previous example, of course, in any actual classification, prior probabilities should also be included.) Figure 5–9 also shows a line labeled *discriminant scores*. This line shows distances from the dividing line. The bell-shaped

FIGURE 5–8

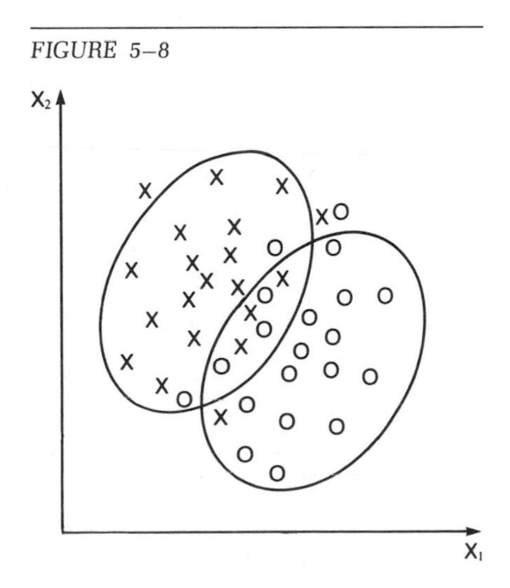

FIGURE 5–9

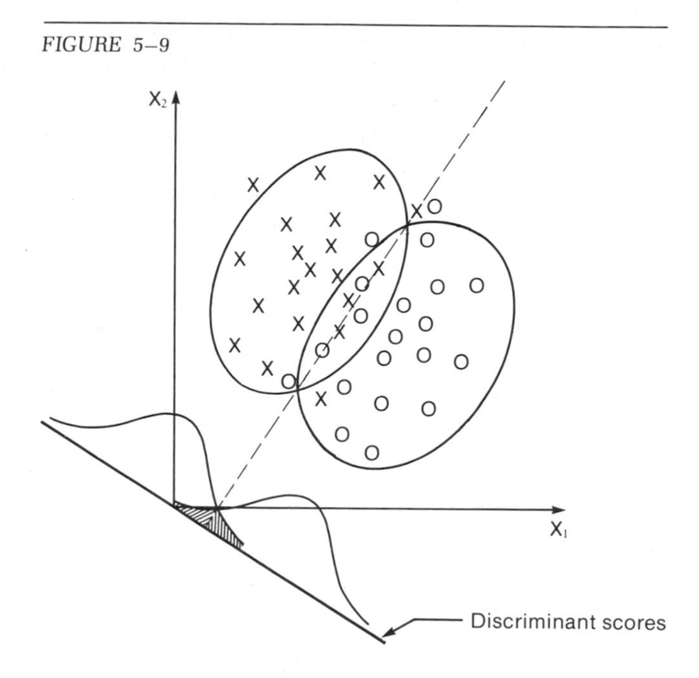

curves show the distributions of such distances that are expected, with the multivariate normal assumption, for Group I and Group II. The shaded region under those curves shows the probabilities of mistakes when classification is done on the basis of likelihoods only. It is entirely analogous to the shaded region in Figure 5–2. The separation for Figure 5–9 is the fraction of each probability distribution that falls on its own side of the dividing line. The separation defined for this figure is just like that in the earlier example.

The output of a computer program for discriminant analysis would provide the separation and a confusion matrix. It would also provide some description of how to find the discriminant scores. The choice as to which such descriptive information to provide varies from program to program. One common possibility is to give an equation defining the discriminant scores (the distances from the dividing line or values proportional to those distances). This equation is called the discriminant function. It turns out that this function is a linear function of X_1 and X_2:

$$a_0 + a_2X_1 + a_2X_2$$

It also turns out that the value of the discriminant function for any point is equal to the logarithm of the probability that the point comes from one of the groups divided by the probability that it comes from the other (with only likelihood information considered). Since it gives logarithms of ratios of probabilities, the function is said to give the log odds of one group over the other.

In interpreting the results of discriminant analysis, users often examine the coefficients of the discriminant function to see which variable is weighted more heavily in discriminating between groups. It is often useful to examine the standardized coefficients—those that would be applied to standardized values of X_1 and X_2. Such a procedure is intended to eliminate effects of scale differences for the discriminator variables.

EXAMPLE ON DISTRIBUTION METHODS

Table 5–3 gives data on a set of industrial products. For each product, the identifier variable indicates whether its method of distribution is direct or through distributors. The table lists two possible discriminator variables. The first is an estimate of the ratio of the size of an average customer order to the size of an average production run for the product. Here "average customer order" means that amount of that particular product ordered at one time and does not include the amount of any

TABLE 5–3

Observation	*Product*	*Distribution (1 = distributor 0 = direct)*	*Ratio (order size/ production run)*	*Service (nontechnical service)*
1	Grinding wheels (standard)	1	.3	3
2	Cutting tools (drill bits, etc.)	1	.1	3
3	Pipe fittings (stainless steel)	1	.2	5
4	Electrical cable (sold to building contractors)	1	.3	7
5	Fine white paper (sold to printers)	1	.1	6
6	Diesel engines (to owner-operators)	1	.1	8
7	Over-the-road tractors	1	.4	6
8	Automotive painting systems	0	1.0	6
9	X-ray systems (to hospitals)	0	.6	4
10	Disposable syringes	1	.1	3
11	Jet engines	0	.6	5
12	Photo identification badge machines	0	.8	3
13	Small business computers	0	.2	6
14	Large PBXs (telephone equipment)	0	.8	1
15	Commercial airframes	0	1.0	3
16	Titanium chemical reactor vessels	0	1.0	8
17	Work gloves	1	.1	2
18	Rock salt	0	.3	4
19	Isopropanol (to the coatings industry)	1	.3	4
20	Engineered plastics	0	.3	7

other products that might be included on the same invoice. This variable is intended to capture at least parts of a number of aspects of the product, including: the degree to which the product is customized for individual customers, the economies of scale in its manufacture, and the levels of inventories maintained for the product. It is expected that more customization, less economies of scale, and lower inventory levels will all be associated (though not perfectly associated) with higher values of the first discriminator variable.

The second variable is intended to estimate the required level of nontechnical service. Such service aspects as frequent delivery are included; such aspects as engineering problem solving are not. The scale for this variable runs from 1 (very low service level) to 8 (very high level).

We assume that the products in this initial database are representative industrial products and that their methods of distribution are considered appropriate to them. The values of the two variables are not known exactly for these or other products, so the values in the table were obtained as judgments from someone knowledgeable about sales and industrial marketing.

Before proceeding to consider results for these data, it is useful to emphasize two points about the data collection. First, it is important that the values for the discriminator variables be determined for the input data in exactly the same way they will be obtained for other observations to be classified with the results of the discriminant analysis. (This condition is necessary when we use any of the multivariate techniques on one set of data and then apply the results to additional observations.) Thus, we assume that for future observations we will obtain judgmental values for the two discriminator variables in the same way from the same expert.

Second, it is important that the products in the database be sufficiently like those for which future classification will be performed. This requirement may seem obvious, but the principle behind it is often violated in discriminant analysis. Problems of this sort often arise in credit scoring. Analysts often fit discriminant functions to individuals who were granted credit in the past. They then apply the resulting classification procedures to a wide range of future applicants. The problem is that the individuals previously granted credit had presumably passed some less formal screening procedure. Hence, they are not at all representative of the broader range of individuals who apply for credit (as opposed to those who were granted such credit). The resulting classification procedure may (or may not)[5] be very good for predict-

[5]For the procedure to work well even on those who pass the screen, the data must satisfy to an acceptable level the usual distributional and other conditions of discriminant analysis—and the discriminator variables must be effective ones.

ing credit behavior of those who pass the same screen that was used in the past, but there is no guarantee that the procedure will be a satisfactory replacement for such a screen (in other words, that it will work on the wider group of applicants). Only a procedure fit to such a wider group can sensibly be expected to work for a wider group.

One aspect of the data set may bother the reader and is worth comment at this point. The discriminant analysis model assumes that the discriminator variables follow a multivariate normal distribution. Also, users generally assume equal variance structures in the groups. Often, users are not at all sure about the variance structure assumption. Figure 5–10 shows a plot of the industrial-products data. The figure does not present strong support for the variance structure assumption.[6]

The user may also wonder whether the discriminator variables are even individually (marginally) approximately normally distributed in each of the two populations. In other cases, users want to use discriminator variables which are very clearly *not* even approximately normally

FIGURE 5–10

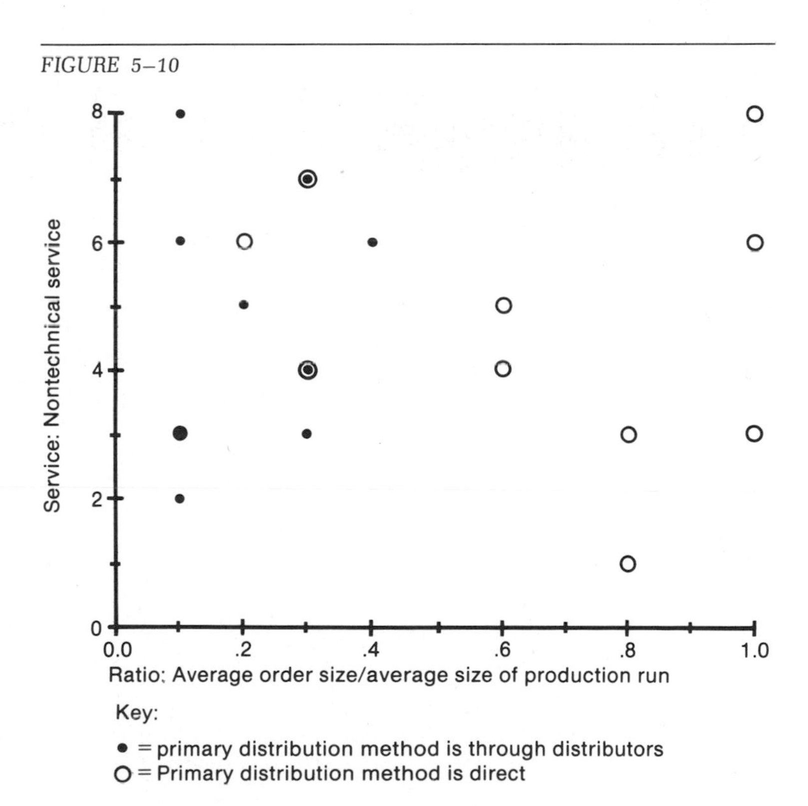

[6]The figure shows more dispersion *in the sample* of discriminator variables for the direct-distribution group. That fact suggests, but does not prove, more dispersion in the underlying direct-distribution *population*.

distributed. For example, they may think that a dummy variable would be a good discriminator. Yet a dummy variable is certainly not normally distributed and hence cannot fit the assumptions of the discriminant model.

It turns out in practice that the discriminant analysis model is surprisingly robust. In other words, the discriminant procedure is found to work well even when its assumptions are not met. Investigators regularly use the technique when they do *not* believe that their groups have essentially the same variance structures. (Though the technique should not be used when variance structures are extremely different.) Further, investigators regularly use discriminator variables which are not normally distributed, even marginally; dummy variables are sometimes included. Even in such cases, discriminant analysis is found to give useful results. We do not have theoretical proofs that it should do so, yet practice indicates that it does.

The industrial distribution data were first subjected to a discriminant analysis using only ratio (order size/production-run size) as a discriminator. Equal variances were assumed. The discriminant analysis program fit the following equation for the log odds:

$$\log\left\{\frac{\text{probability of distributor}}{\text{probability of direct}}\right\} = 3.618 - 8.415 * \text{ratio}$$

For every increase of .1 in the ratio, the log odds in favor of the distributor method decreases .8415. The computer program separately reported a separation of 84 percent: 84 percent of the estimated distribution of ratio values for direct-method products falls above the dividing value.[7] The same fraction of the estimated distribution of ratio values for distributor-method products falls below the dividing value.

Table 5–4 gives the confusion matrix. It shows that all the distributor products, but only 7 of the 10 direct products, were classified correctly with the discriminant function.

TABLE 5–4

		"Predicted" Method		
		Direct	*Distributor*	*Total*
True Method	*Direct*	7	3	10
	Distributor	0	10	10
	Total	7	13	20

[7]The dividing values turns out to be approximately .43.

TABLE 5–5

Observation	Ratio	Actual Distribution Method	Posterior Probabilities: Distributor	Posterior Probabilities: Direct	Predicted Distribution Method	Correct?
1	.3	1	.728	.272	1	Yes
2	.1	1	.911	.089	1	Yes
3	.2	1	.845	.155	1	Yes
4	.3	1	.728	.272	1	Yes
5	.1	1	.911	.089	1	Yes
6	.1	1	.911	.089	1	Yes
7	.4	1	.557	.443	1	Yes
8	1.0	0	.031	.969	0	Yes
9	.6	0	.218	.782	0	Yes
10	.1	1	.911	.089	1	Yes
11	.6	0	.218	.782	0	Yes
12	.8	0	.072	.928	0	Yes
13	.2	0	.845	.155	1	No
14	.8	0	.072	.928	0	Yes
15	1.0	0	.031	.969	0	Yes
16	1.0	0	.031	.969	0	Yes
17	.1	1	.911	.089	1	Yes
18	.3	0	.728	.272	1	No
19	.3	1	.728	.272	1	Yes
20	.3	0	.728	.272	1	No

Table 5–5 shows how the entries in Table 5–4 would be found. It gives the posterior probabilities found by a computer program based on the assumption that the prior probabilities are equal (as in the database).[8] It then gives the classification choice when the observations are classified on the basis of the posterior probabilities.

These discriminant results can now be used to predict the distribution method for other products. Assume that the prior probabilities are equal. Table 5–6 shows the posterior probabilities of distribution method found by the discriminant program for products with ratios of .3, .5, .7, and .9. The first would be predicted to be a distributor product and the rest to be direct ones. If a product with a .5 ratio were in fact

[8]In finding the posterior probabilities, this particular discriminant analysis program considers:

1. The log odds as given by the discriminant function.
2. Additional consideration of uncertainty about the exact normal distributions and the values of the discriminant coefficients (since these values are not known precisely).

These calculations are not at all easily replicated by hand.

TABLE 5–6

Ratio	Posterior Probabilities: Distributor	Posterior Probabilities: Direct	Predicted Distribution Method
.3	.728	.272	1
.5	.369	.631	0
.7	.123	.877	0
.9	.045	.955	0

distributed through distributors rather than directly, as predicted by the model, that product might be a good candidate for further attention from the consultants in the example.

In a separate run, the program was instructed not to assume equal variances in the two populations. The resultant confusion matrix was exactly the same as the one in Table 5–4. The posterior probabilities for other products with this alternative model are given in Table 5–7. Note that they differ from those in Table 5–6, although the classifications of the new products have not changed.

Next, the program was instructed to use both discriminator variables in fitting a discriminant function, using an assumption of equal variances and covariances. The equation was:

$$\log\left\{\frac{\text{probability of distributor}}{\text{probability of direct}}\right\} = 3.902 - .05763 * \text{service} - 8.445 * \text{ratio}$$

The separation was 84 percent. Notice that the sign of the coefficient of the ratio variable is as expected: the higher the ratio, the lower the odds of the distributor method. The sign of the service variable is not as expected, however. We would expect higher values of that variable to increase rather than decrease the odds of the distributor method. Note,

TABLE 5–7

Ratio	Posterior Probabilities: Distributor	Posterior Probabilities: Direct	Predicted Distribution Method
.3	.779	.221	1
.5	.186	.814	0
.7	.013	.987	0
.9	.002	.998	0

TABLE 5–8

Ratio	Service	Posterior Probabilities: Distributor	Posterior Probabilities: Direct	Predicted Distribution Method
.5	1	.422	.578	direct
.7	4	.116	.884	direct
.8	1	.097	.903	direct

in addition, that the ratio variable is weighted considerably more heavily in the discriminant function than is the service variable.[9]

It would, of course, be possible to use the two-discriminator model. The confusion matrix turns out to be the same as that in Table 5–4. Table 5–8 gives the posterior probabilities for several combinations of values for the discriminator variables. It does not seem sensible to use the results of this model, however, since the sign of one coefficient doesn't make sense. Instead, it would seem preferable either to use the earlier single-discriminator model or else to search for some other second discriminator that gave more sensible results.

DISCRIMINANT ANALYSIS WITH ADDITIONAL DISCRIMINATORS OR GROUPS

With more than two discriminators, the discriminant analysis procedure is entirely analogous to that described above. The only difference is that the discriminant function will now have three or more discriminator variables with their respective coefficients.

With more than two groups, the process becomes a bit more complex. The basic procedure is still to fit multivariate normal distributions for each of the groups and then to use the results in classification. The ideas of separation and of the confusion matrix remain what they were above. Now, however, information must be found to separate all of the groups from one another. Different programs use different ways of describing the results, and users of discriminant analysis must determine the method of presentation of the particular program they use. One simple choice is to print a function for each pair of groups, giving the log odds of one of those groups over the other. The function will be

[9] The standardized coefficients are −.1186 for service and −1.975 for ratio. These coefficients apply to standardized versions of the variables.

a linear one,[10] just as was the single discriminator function in the case of only two groups. An example of a problem with three groups might involve the method of compensation for salespeople selling industrial products. The groups would be commission, salary, and mixed. Discriminator variables might be selected to reflect complexity of the selling task, length of selling time per sale, and other measures. There would then be one equation for the log odds of salary over commission:

$$\log\left\{\frac{\text{probability of salary}}{\text{probability of commission}}\right\} = b_0 + b_1 * \text{complexity} + b_2 * \text{time}$$

There would be another equation for the log odds of mixed over salary and a third for the log odds of mixed over commission:

$$\log\left\{\frac{\text{probability of mixed}}{\text{probability of salary}}\right\} = b_0' + b_1' * \text{complexity} + b_2' * \text{time}$$

$$\log\left\{\frac{\text{probability of mixed}}{\text{probability of commission}}\right\} = b_0'' + b_1'' * \text{complexity} + b_2'' * \text{time}$$

(The primes and double primes on the coefficients in these equations are intended to convey the idea that, although all three equations have the same form, they will in general have different specific sets of coefficient values.)

For such a problem, the discriminant analysis program would provide a three by three confusion matrix. It would produce a set of three posterior probabilities for each new set of values for the discriminator variables; there would be one posterior probability per group. Similarly, discriminant analysis can be applied to problems involving larger numbers of groups.

---

[10]Under the assumption of equal variance structures.

CHAPTER 6

Principal Components Analysis

Principal components analysis is a mathematical method for expressing a set of information (in the form of a set of data) in an alternative form. The method is a purely mathematical one; it does not involve an underlying model of the world. Rather, it simply restates a particular set of data in one particular way.[1] Principal components analysis is often called a technique for data reduction rather than data restatement, because the specific form of restatement used is sometimes considered to allow data reduction. In other words, the restatement may allow investigators to capture most of the information in the original data set considerably more succinctly in a reduced data set. For purposes of explanation it is best to begin a discussion of principal components analysis as a restatement device. After that discussion, the possible use of the tool for data reduction will be easier to understand.

GEOMETRIC DEFINITION

One way of considering principal components analysis is in terms of the geometric representation of data. We begin with this approach. So that the geometry can be depicted easily, we will consider two dimensions to start. Table 6–1 contains a listing of a set of 12 observations in terms of two variables, X_1 and X_2. The observations have been labeled with the letters A through L. Figure 6–1 is a geometric plot of these observations, with the letter labels included.

We can represent points A through L in other ways as well. To begin, consider Figure 6–2, in which the 12 observations have been

[1]In particular, principal components analysis does not involve a dependent variable. All of the variables in the input data serve similar roles in the technique; none is singled out to be explained by the others.

TABLE 6–1

Point (observation)	X_1	X_2
A	8	4
B	5	6
C	5	2
D	3	−2
E	2	4
F	1	−3
G	0	0
H	−1	2
I	−3	−1
J	−5	−5
K	−6	−1
L	−8	−4

FIGURE 6–1

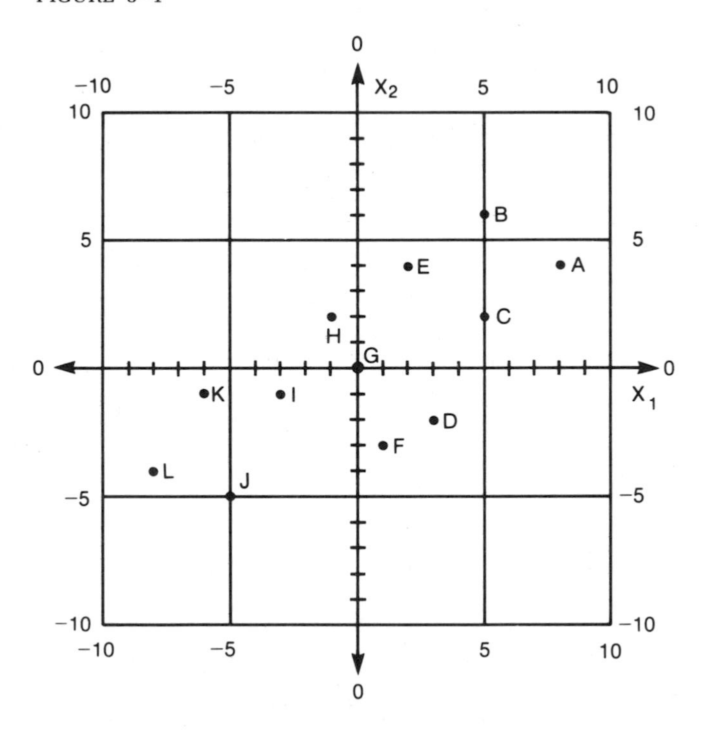

plotted without any axes shown. (Figure 6–1 is identical to Figure 6–2 except for the inclusion of the axes.) Figure 6–2 can be thought of as an abstraction from Figure 6–1 of the information about the relative positions of the 12 points, without the use of the axis information.

There is no reason why the information in Figure 6–2 cannot be interpreted relative to some set of axes other than the ones used in

FIGURE 6–2

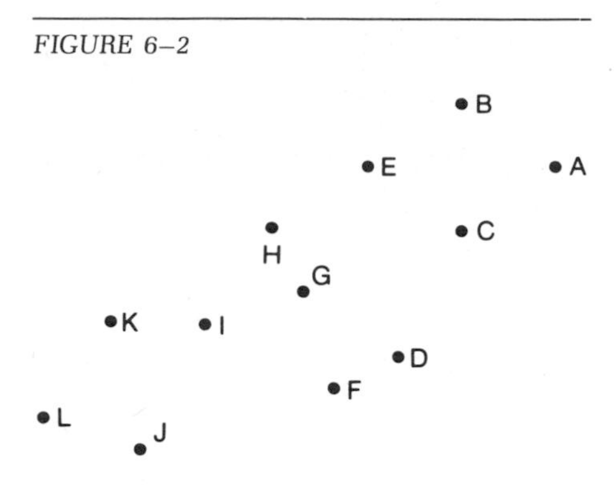

Figure 6–1. For example, in Figure 6–3 the 12 points are in the same positions on the page as they are in Figure 6–1 or Figure 6–2. (In fact, the points for Figures 6–2 and 6–3 were obtained by tracing from Figure 6–1.) In the third figure, however, a different set of axes (or *coordinate axes*) has been used. The observations can be described by their coordinates along these new axes just as well as they can be described along the axes in Figure 6–1. For example, point G remains the origin: it has coordinates 0 and 0 along the X_1 and X_2 axes in Figure 6–1 and coordinates 0 and 0 along the $X_1{}^*$ and $X_2{}^*$ axes in Figure 6–3. The other points all have different coordinates relative to the new axes. Point A has new coordinates of approximately 8.7 in the $X_1{}^*$ direction and 2.0 in the $X_2{}^*$ direction. Similarly, we can read the coordinates of the 10 other points.

Even if we consider only sets of axes which leave the point G at the origin, there are many other possible choices. For example, Figure 6–4

FIGURE 6–3

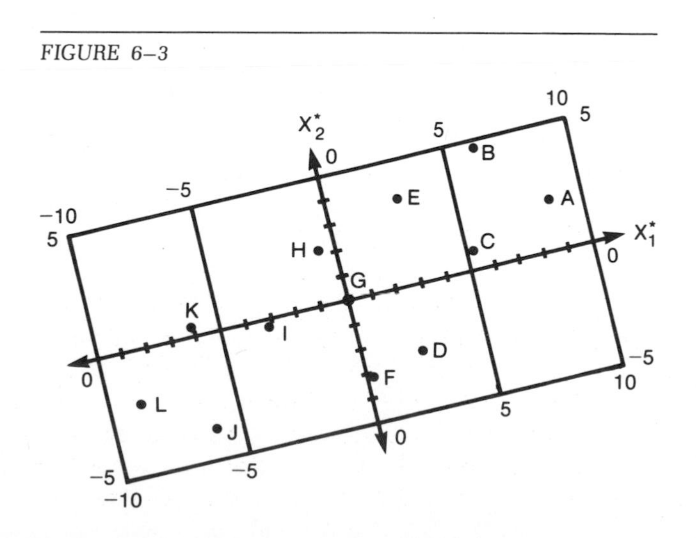

FIGURE 6–4

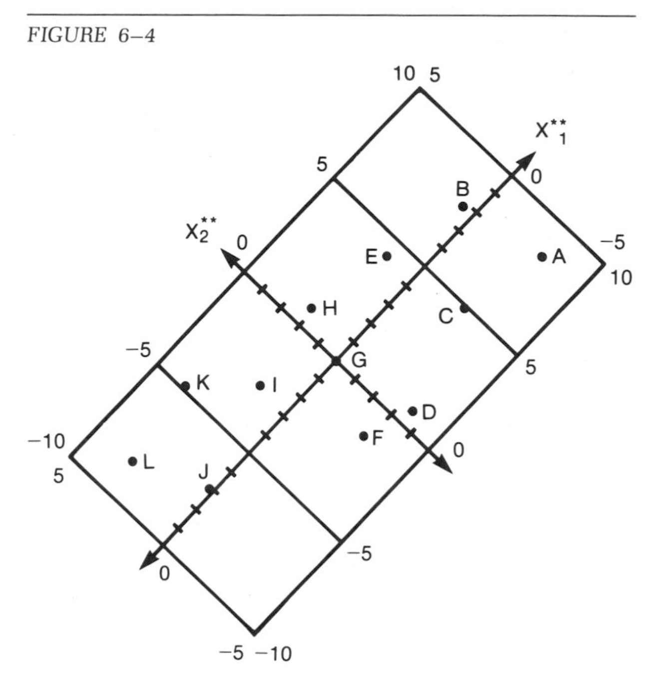

shows another set of axes. Again, the points in the figure were obtained by tracing from Figure 6–1, so that the relative configuration of the points remains what it was in earlier figures. In Figure 6–4, point A has coordinates of approximately 8.4 in the $X_1{}^{**}$ direction and -3.1 in the $X_2{}^{**}$ direction. We can also find coordinates of the other points relative to these new axes.

In geometric terms, principal components are defined as a special set of coordinate axes obtained from the original set used in describing a set of data. In two dimensions, we start with a set of data expressed relative to two axes. An example is given in Table 6–1 and Figure 6–1.[2] The first principal component of these data is then defined as that new axis (or direction) which explains as much as possible of the variability in the relative placement of the original set of data points. What this definition means is that the first principal component is that axis which is nearest to the set of data points in a squared-distance sense. Figure 6–5 shows the 12 data points from Figure 6–2 together with one candidate axis. The distances from each of the points to the candidate axis are shown in the figure; the distances are the lengths of the perpendicular lines from the points to the candidate axis. The fit of this particular candidate axis to the data points is measured by the sum of the squared distances. The first principal component is that axis, from

[2]Customarily in principal components analysis, investigators use variables (transformed if necessary) with zero means. The variables in Table 6–1 do not have zero means.

FIGURE 6–5

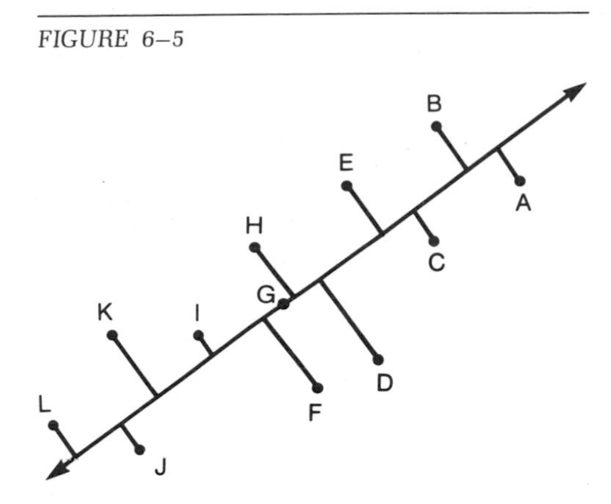

among all the possible candidates, for which the sum of the squared distances is smallest. The second principal component, or second axis, is then taken perpendicular (at right angles) to the first.

Effects of Scale

Before leaving the geometric development of principal components in two dimensions, we now turn to another example with which we can consider the effects on the results of using different scales for measuring the original variables (or axes). Table 6–2 contains a listing of six observations described by two variables X_1 and X_2. Figure 6–6 shows a plot of this information. The first principal component for these data appears to be very close to the horizontal (the X_1 direction).

TABLE 6–2

Point	X_1	X_2
M	−3	.2
N	−1.5	−.3
O	0	.1
P	1.5	.3
Q	2	−.1
R	3	.08

FIGURE 6–6

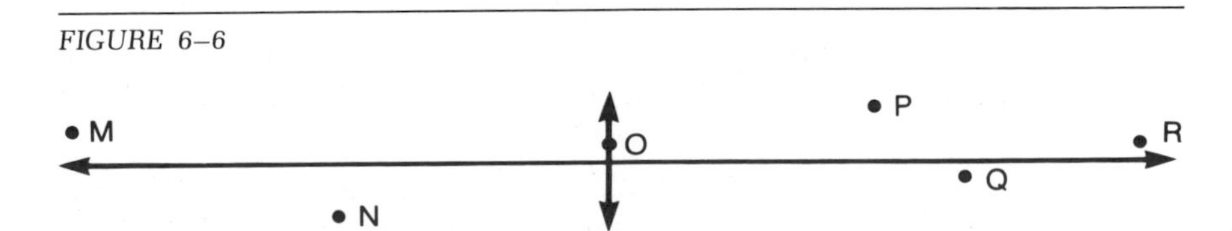

TABLE 6–3

Point	X_1^*	X_2^*
M	−.3	2
N	−.15	−3
O	0	1
P	.15	3
Q	.2	−1
R	.3	.8

FIGURE 6–7

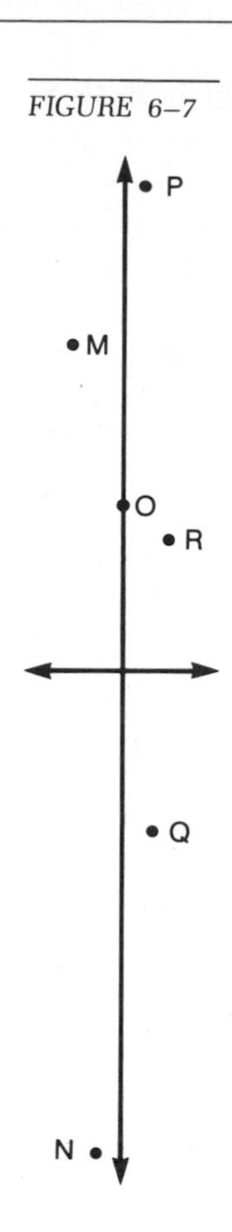

Now, suppose that another investigator collected basically the same information as that given in Table 6–2 but used a different scale of measurement. In particular, assume that this investigator used variables X_1^* and X_2^* where one unit of X_1^* is equivalent to 10 units of X_1 and where one unit of X_2^* is equivalent to $\frac{1}{10}$ unit of X_2. Table 6–3 lists the six data points in terms of X_1^* and X_2^*, and Figure 6–7 plots the points. The first principal component now appears to be very close to the vertical (the X_2^* direction).

Because the effects of the scales on which variables are measured can be so important in determining principal components, it is common for analysts who use this technique to standardize their variables. In other words, they transform the variables so that each has variance 1 (and, usually, mean 0) before determining the components. The motivation behind standardizing is a desire to give each variable equal importance. As will be discussed further below when we consider data reduction, it is not at all clear in many cases that the assumption involved in using standardized variables is an appropriate one in principal components. Sometimes one variable or the variation in the values of one variable is simply more important than another. On the other hand, it is appropriate to know that standardization is extremely common in the use of principal components analysis.

The geometric definition of principal components can, in theory, easily be extended to more than two dimensions. Given an initial set of observations expressed relative to some number m of dimensions (variables), we define the first principal component as that direction axis from which the sum of the squared distances is smallest. (That axis explains as much as possible of the variation or placement of the points.) The second principal component is then defined as the axis, from among all possible axes perpendicular to the first component, with the smallest sum of squared distances. (It explains as much as possible of the remaining dispersion or placement of the points.) Similarly, the third principal component is that axis, from among all those perpendicular to the first two, with the smallest sum of squared distances. (It explains as much as possible of the remaining dispersion of the points.) Finally, after $m - 1$ axes have been selected in this manner, there will exist only one possible axis that is perpendicular to all of the axes chosen thus far. That unique perpendicular is defined as the mth principal component.

ALGEBRAIC DEFINITION

In practice, while the geometric definition is quite correct in more than two dimensions, it is difficult for most people to deal with geometry in more than two or three dimensions. Consequently, we generally turn to an equivalent algebraic definition of principal components.

To introduce the algebraic definition, we return to the set of data given in Table 6–1 above. The information contained in the values in that table for variables X_1 and X_2 can be expressed in other ways. For example, if we define new variables $X_1{}^*$ and $X_2{}^*$ by

$$X_1{}^* = X_1 + X_2$$
$$X_2{}^* = X_2 - X_1$$

then the values of $X_1{}^*$ and $X_2{}^*$ for each observation are as shown in Table 6–4.

The values of $X_1{}^*$ and $X_2{}^*$ can be found from the X_1 and X_2 values, as shown in the table. It is also possible to reverse the procedure and to find the X_1 and X_2 values from the $X_1{}^*$ and $X_2{}^*$ values.

TABLE 6–4

Point	X_1	X_2	$X_1^* = X_1 + X_2$	$X_2^* = X_2 - X_1$
A	8	4	12	−4
B	5	6	11	1
C	5	2	7	−3
D	3	−2	1	−5
E	2	4	6	2
F	1	−3	−2	−4
G	0	0	0	0
H	−1	2	1	3
I	−3	−1	−4	2
J	−5	−5	−10	0
K	−6	−1	−7	5
L	−8	−4	−12	4

In fact

$$X_2 = .5(X_1^* + X_2^*)$$
$$X_1 = .5(X_1^* - X_2^*)$$

Thus, the information in Table 6–1 can be presented in terms of X_1 and X_2 or it can be presented just as well in terms of X_1^* and X_2^*. The two sets of variables are simply restatements of one another. Similarly, there are many other sets of variables that could be used.

The principal components of a set of data are one particular set of variables that can be used for restatement of the original variables. To understand one part of the definition of principal components, it is useful to consider a plot of the X_1^* and X_2^* variables that are listed in Table 6–4. Figure 6–8 presents such a plot. Comparison of Figures 6–1

FIGURE 6–8

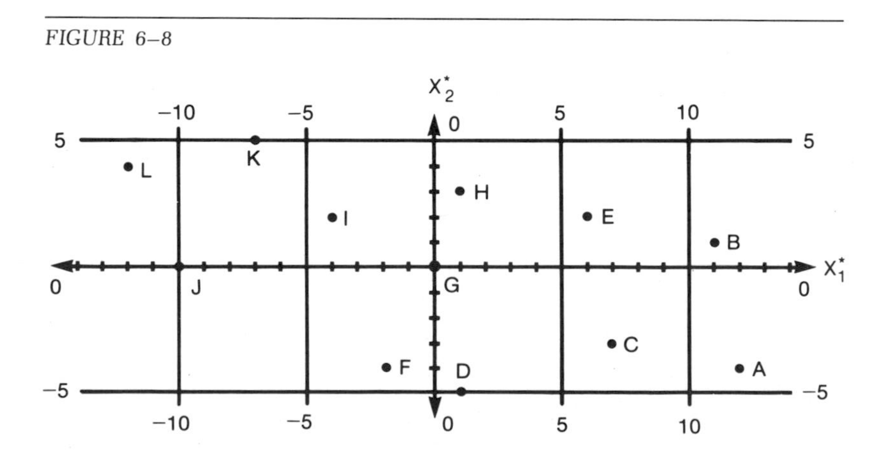

and 6–8 with the aid of a ruler shows that the scale of Figure 6–8 is larger. Even though the two figures were constructed with the same length per unit for each of the axes, the distances between points in Figure 6–8 are larger than the distances between the corresponding points in Figure 6–1. In fact, the scaling in Figure 6–8 is quite arbitrary. New variables defined as $2X_1^*$ and $2X_2^*$ would be completely capable of conveying the information in the original data set, as would many other rescalings of the variables. To identify the principal components uniquely from among all possible restatements of the original variables, we include in the definition of principal components a statement specifying the scale of the new variables. In addition, we impose the more important condition that the first principal component explain as much as possible of the variation in the original data, and so on, much as we did in the geometric definition above.[3] A more complete definition of principal components, in algebraic terms, follows.

Suppose that we have values for some number m of original variables $X_1, X_2, \ldots, X_m$ on each of n observations. Suppose for convenience that each of the original variables has mean 0.[4] Our original database can be described as shown in Figure 6–9.

The first principal component of the set of data is the new variable ξ_1 satisfying the following conditions:

1. ξ_1 is a linear function of the original variables $X_1, X_2, \ldots, X_m$. In other words,

$$\xi_1 = w_{11}X_1 + w_{12}X_2 + \ldots + w_{1m}X_m$$

where w_{11}, w_{12}, and w_{1m} are constants (weights) defining the linear function.

2. $$w_{11}^2 + w_{12}^2 + \ldots + w_{1m}^2 = 1$$

(This statement is the one that specifies formally the scale of the new variable ξ_1.)[5]

3. Of all linear functions of the original variables satisfying the first two conditions, ξ_1 has the following property: Find the value of ξ_1 on each observation of the data. Find the variance of the values. The linear function defined as the first principal component is the

[3]In the geometric presentation, it is most natural to think in terms of minimizing a sum of squared distances. In the algebraic version, we think instead of maximizing the explained variation. The two conditions are equivalent.

[4]This condition is not really necessary but it is customary; it simplifies the definition that follows, and it is not restrictive. (We can always make a variable have mean 0 by adding or subtracting an appropriate amount from its value on each observation.)

[5]Other choices are possible for the scaling condition. The one given here is the most common one.

FIGURE 6–9

	X_1	X_2	X_3	• • •	X_m
Observation 1				• • •	
Observation 2				• • •	
⋮	⋮	⋮	⋮	⋮ ⋮ ⋮	⋮
Observation n				• • •	

one that maximizes this variance. (This statement is the formal condition that the first principal component explain as much as possible of the variation of the original data points. It is equivalent to the geometric condition that the sum of the squared distances from the input observations to the first principal component be a minimum.)

Once the first principal component is defined, the second principal component, ξ_2, is specified by the following conditions:

1. ξ_2 is a linear function of the original variables.

$$\xi_2 = w_{21}X_1 + w_{22}X_2 + \ldots + w_{2m}X_m$$

where $w_{21}, w_{22}, \ldots, w_{2m}$ are constants (weights) defining the linear function.

2.

$$w_{21}^2 + w_{22}^2 + \ldots + w_{2m}^2 = 1$$

(a scaling condition)

3.

$$w_{11}w_{21} + \ldots + w_{1m}w_{2m} = 0$$

(This statement is the algebraic equivalent of the geometric requirement that the two new axes ξ_1 and ξ_2 be perpendicular.)

4. Suppose that we find the values for the two new variables for each observation in the database. These variables must be uncorrelated.

5. Subject to all of these conditions, ξ_2 is selected to have the maximum variance. (This condition is equivalent to the requirement that ξ_2 explain as much as possible of the remaining variation—and also to be the requirement that the sum of the squared distances from ξ_2 be as small as possible.)

Proceeding in this way, we find m principal components (as many components as there were variables in the original set).[6] At each stage we select a new component that is a linear function of the original variables, is perpendicular to the previously selected components, with values uncorrelated with the previous sets of values, and which explains as much as possible of the remaining variability in the data. When we are through, we will have a set of original variables X_1, $X_2, \ldots, X_m$ and a set of principal components $\xi_1, \xi_2, \ldots, \xi_m$, each defined on each observation of the database. As was true for the example in Table 6–4, it will be possible to express either set of variables in terms of the other set. In other words, the definition just given specifies that each ξ must be a linear function of $X_1, \ldots, X_m$. After we have defined the $\xi_1, \xi_2, \ldots, \xi_m$, it will be equally possible to express the original X variables as linear functions of the ξs where we might call the weights $\alpha_{11}, \alpha_{12}, \ldots$. The expression is sketched below:

$$X_1 = \alpha_{11}\xi_1 + \alpha_{12}\xi_2 + \ldots + \alpha_{1m}\xi_m$$
$$X_2 = \alpha_{21}\xi_1 + \alpha_{22}\xi_2 + \ldots + \alpha_{2m}\xi_m$$
$$\vdots$$
$$X_m = \alpha_{m1}\xi_1 + \alpha_{m2}\xi_2 + \ldots + \alpha_{mm}\xi_m$$

The variances of the original variables will have been completely accounted for by the principal components. In other words, suppose that we find the variance in the database of the first variable X_1 and that we call this value V_1. Similarly, suppose $V_2, V_3, \ldots, V_m$ are the variances of the other original variables $X_2, X_3, \ldots, X_m$. Define VAR as the sum.

$$VAR = V_1 + V_2 + \ldots + V_m$$

ξ_1 is selected to explain as much as possible of VAR. ξ_2 is selected to explain as much as possible of what remains, and so on. Suppose we call the variance of the ξ_1 values U_1, the variance of the ξ_2 values $U_2, \ldots$, and the variance of the ξ_m values U_m. When we have determined all of the principal components we will find

$$V_1 + V_2 + \ldots + V_m = VAR = U_1 + U_2 + \ldots + U_m$$

[6]To be precisely correct, we should note that we will find m principal components only if none of the original variables is a linear combination of others of those original variables.

Also, we note that there must be some convention about the signs of the weights. (In geometric terms, we had to specify in Figure 6–3, for example, whether $X_1{}^*$ increased from left to right or from right to left.) For example, we can require that the sum of the positive weights in any component be at least as large as minus the sum of the negative ones.

In determining principal components, *VAR* is the target. ξ_1 is selected so that it contains as much as possible of the variability: in other words, so that U_1 is as large as possible. ξ_2 is then selected to give as large as possible a U_2, and so on.

Because the target to be explained is the sum of the variances of the individual variables, it should be clear that original variables with large variances will be more important in determining principal components than will original variables with smaller variances. (Such variables will contribute more to the target sum.) In attempts to treat all variables "equally," investigators routinely standardize the original variables, transforming them to have means of 0 and variances of 1. The advisability of such measures can best be considered in the context of applications of principal components later in the chapter.

The actual calculation of the principal components of a set of data is a straightforward, unambiguous matter for a computer. Consequently, we need not consider the details, except to note that the calculations are performed using the correlations among the original variables. (Hence, programs for a principal components analysis will require a matrix of correlation coefficients.) Instead, we can proceed to consider possible applications of the technique.

PRINCIPAL COMPONENTS FOR DATA REDUCTION

As defined above, principal components analysis is simply a method for restating the information in an original set of variables in terms of a new set of variables with successively smaller variances. In practice, however, the technique is often used for data reduction. The argument is that since the first principal component is selected to contain as much as possible of the variability of the original set of variables, perhaps that component could be used, alone, to summarize much of the useful information in the data. The component is thought of as an index. The idea is to use a single index (or, at most, a few indices) in place of the full original set of variables. Hence the term *data reduction*.

An example is provided by the work of Kendall (1939) in which he collected data for 48 counties in Great Britain on the yields (per acre) of 10 crops.[7] The observations were the 48 counties. The original variables were yields for the 10 crops: wheat, barley, oats, beans, peas, potatoes, turnips, mangolds (a type of beets), temporary hay, and permanent hay. The variables were standardized so that each had variance 1. Hence, the target sum of variances to be explained was 10.

---

[7]Reproduced in: Maurice Kendall, *Multivariate Analysis* (New York: Hafner Press, 1975).

Kendall found that the first principal component was

$$
\begin{aligned}
\xi_1 = {} & .39 * \text{wheat} && + .37 * \text{barley} \\
& + .39 * \text{oats} && + .27 * \text{beans} \\
& + .22 * \text{peas} && + .30 * \text{potatoes} \\
& + .32 * \text{turnips} && + .26 * \text{mangolds} \\
& + .24 * \text{temporary hay} && + .34 * \text{permanent hay}
\end{aligned}
$$

That component had variance 4.76; in other words, it explained 47.6 percent of the sum of the variances of the original variables. Kendall suggested using ξ_1 as a single index of productivity, in place of the 10 original yield variables. (He could also have considered using multiple principal components as indices.)

In a second example, Feeney and Hester (1967) collected data on stock prices from the first quarter of 1951 through the second quarter of 1963 (50 observations).[8] They included 30 stocks in their sample. The weights defining the first and second principal components (ξ_1 and ξ_2) in terms of the original variables are shown in Table 6–5.[9] ξ_1 is suggested as a general index of the stock market; it accounts for 75.76 percent of the variability of the 30 stocks. ξ_2 is suggested as more industry-specific. (It seems to weight retailers and some consumer goods firms heavily in the negative direction and to weight some of the industrial producers positively.) ξ_2 accounts for an additional 13.93 percent of the original variance.

While the use of principal components to define indices may seem appealing, it may well be that the real value in the concept is the basic idea of an index rather than the use of principal components. It isn't clear that a better index of productivity in the first example above couldn't be defined by considering the importances of the different crops to be measured in some other way (such as their total cash value, perhaps) rather than by using the variances. (Variance is not a particularly intuitive concept.) These ideas can be explored further with a specific example.

Suppose that we want to develop an index for the cost of food in major U.S. cities and that we have collected data for March 1973. Table 6–6 shows some of the data. It would certainly be possible to include more variables (foods), but the five in the table will serve for illustrative purposes.

Table 6–7 shows the variances of the five food-price variables. It also gives the total of the variances and the fractions of the total vari-

[8] Described in: S. James Press, *Applied Multivariate Analysis* (New York: Holt, Rinehart & Winston, 1972).

[9] The sum of the squared weights for ξ_1 is considerably more than 1. The investigators apparently used a scaling convention different from the one described in this chapter.

Table 6–5

Stock	*Weight for* ξ_1	*Weight for* ξ_2
Allied Chemical	.137	.039
Alcoa	.405	.308
American Can	.067	.035
AT&T	.361	−.272
American Tobacco	.123	−.081
Anaconda	.101	.168
Bethlehem	.217	.104
Chrysler	−.042	.025
Du Pont	.951	.217
Eastman Kodak	.587	−.231
General Electric	.376	.047
General Foods	.391	−.254
General Motors	.236	.009
Goodyear	.222	−.012
International Harvester	.123	−.051
International Nickel	.297	−.061
International Paper	.156	.054
Johns-Manville	.150	−.004
Owens-Illinois	.374	−.012
Procter & Gamble	.342	−.238
Sears	.319	−.185
Standard Oil (California)	.195	.007
Esso	.198	.068
Swift	.037	.019
Texaco	.251	−.090
Union Carbide	.422	.131
United Aircraft	.186	.155
U.S. Steel	.407	.136
Westinghouse	.145	.004
Woolworth	.175	−.124

Source: S. James Press, Applied Multivariate Analysis (New York: Holt, Rinehart & Winston, 1972).

ance caused by each of the variables. In other words, the total of the five variances is 356.05. The variance of the bread prices is 6.01, which is 1.7 percent of the total.

Recall that in principal components the aim is to construct a first component that explains as much as possible of the total variance of the variables being considered. In this foods example, the target to explain is 356.05. Because the prices for oranges contribute considerably more to the total variance than do the prices for bread, we expect oranges to be more important in the first principal component, or index. In fact, oranges contribute more than half of the total of the five variances.

Table 6–8 summarizes the results obtained with nonstandardized variables used to determine components. The first principal component is

TABLE 6–6

City	White Bread ¢/pound	Hamburger ¢/pound	Milk ¢/1/2 gal	Oranges ¢/doz	Tomatoes ¢/pound
Atlanta	24.5	94.5	73.9	80.1	41.6
Baltimore	26.5	91.0	67.5	74.6	53.3
Boston	29.7	100.8	61.4	104.0	59.6
Buffalo	22.8	86.6	65.3	118.4	51.2
Chicago	26.7	86.7	62.7	105.9	51.2
Cincinnati	25.3	102.5	63.3	99.3	45.6
Cleveland	22.8	88.8	52.4	110.9	46.8
Dallas	23.3	85.5	62.5	117.9	41.8
Detroit	24.1	93.7	51.5	109.7	52.4
Honolulu	29.3	105.9	80.2	133.2	61.7
Houston	22.3	83.6	67.8	108.6	42.4
Kansas City	26.1	88.9	65.4	100.9	43.2
Los Angeles	26.9	89.3	56.2	82.7	38.4
Milwaukee	20.3	89.6	53.8	111.8	53.9
Minneapolis–St. Paul	24.6	92.2	51.9	106.0	50.7
New York	30.8	110.7	66.0	107.3	62.6
Philadelphia	24.5	92.3	66.7	98.0	61.7
Pittsburgh	26.2	95.4	60.2	117.1	49.3
St. Louis	26.5	92.4	60.8	115.1	46.2
San Diego	25.5	83.7	57.0	92.8	35.4
San Francisco	26.3	87.1	58.3	101.8	41.5
Seattle	22.5	77.7	62.0	91.1	44.9
Washington, D.C.	24.2	93.8	66.0	81.6	46.2

Source: *Taken from Estimated Retail Food Prices by Cities, March 1973*, U.S. Department of Labor, Bureau of Labor Statistics, pp. 1–8.

TABLE 6–7

Variable	Variance	Percentage of Total Variance
Bread	6.01	1.7%
Hamburger	54.60	15.3
Milk	46.21	13.0
Oranges	193.94	54.5
Tomatoes	55.29	15.5
Total =	356.05	100%

$$\xi_1 = .028 * \text{bread} + .200 * \text{hamburger} + .042 * \text{milk} + .939 * \text{oranges} + .276 * \text{tomatoes}$$

This index weights oranges very heavily indeed. Tomatoes and hamburger are weighted less heavily, and bread and milk receive very little weight.

With this index, Baltimore is the least expensive city, followed by Los Angeles, Atlanta, Washington D.C., Seattle, and San Diego. Hono-

TABLE 6–8

Based on Five Nonstandardized Variables

A. *Definitions of the principal components*

ξ_1 = .028 * bread + .200 * hamburger + .042 * milk
+ .939 * oranges + .276 * tomatoes

ξ_2 = .165 * bread + .632 * hamburger + .442 * milk
− .314 * oranges + .528 * tomatoes

ξ_3 = − .021 * bread − .254 * hamburger + .889 * milk
+ .121 * oranges − .361 * tomatoes

ξ_4 = .190 * bread + .659 * hamburger − .108 * milk
+ .069 * oranges − .717 * tomatoes

ξ_5 = .967 * bread − .249 * hamburger − .036 * milk
+ .015 * oranges + .034 * tomatoes

B. *Variance explained by each component*

ξ_1:	209.5	
ξ_2:	87.7	
ξ_3:	36.0	
ξ_4:	19.9	
ξ_5:	2.9	
	356.0	Total variance explained

C. *Values of* ξ_1 *for each city*

City	Value	City	Value
Atlanta	−22.5	Los Angeles	−22.6
Baltimore	−25.3	Milwaukee	8.7
Boston	5.8	Minneapolis–St. Paul	3.0
Buffalo	14.1	New York	11.9
Chicago	2.4	Philadelphia	−.9
Cincinnati	−2.2	Pittsburgh	14.0
Cleveland	5.8	St. Louis	10.7
Dallas	10.8	San Diego	−15.1
Detroit	7.2	San Francisco	−4.2
Honolulu	35.6	Seattle	−15.2
Houston	2.0	Washington, D.C.	−20.3
Kansas City	−3.9		

lulu is by far the most expensive, followed by Buffalo, Pittsburgh, New York, Dallas, and St. Louis.

It is not clear that the price of fresh oranges in March is a more important contributor to food costs then is the price of bread. As noted above, it is common in principal components for investigators to standardize the variables, transforming each to have a variance of 1 (and often a mean of 0). If we did so for the foods data, we would by definition then have a target variance to explain of 5 (1 for each of the variables). In principal components analysis of such standardized data, the variability of bread prices would be considered to be exactly as important as that of orange prices or tomato prices.

Table 6–9 summarizes the results obtained with standardized variables. The first principal component is

$$\xi_1 = .496 * \text{bread} + .576 * \text{hamburger} + .340 * \text{milk} + .225 * \text{oranges} + .506 * \text{tomatoes}$$

Now oranges receive the least weight of the five variables. With this index, Seattle is the least expensive, followed by San Diego, Houston, Cleveland, and Los Angeles. Honolulu is most expensive, followed closely by New York and then by Boston.

The results of running principal components for the standardized data are not necessarily satisfactory, either. The idea of such analysis

TABLE 6–9

Based on Five Standardized Variables

A. *Definitions of the principal components*

ξ_1 = .496 * bread + .576 * hamburger + .340 * milk + .225 * oranges + .506 * tomatoes

ξ_2 = − .309 * bread − .044 * hamburger − .431 * milk + .797 * oranges + .287 * tomatoes

ξ_3 = − .386 * bread − .262 * hamburger + .835 * milk + .292 * oranges − .012 * tomatoes

ξ_4 = .509 * bread − .028 * hamburger + .049 * milk + .479 * oranges − .713 * tomatoes

ξ_5 = .500 * bread − .773 * hamburger − .008 * milk + .006 * oranges + .391 * tomatoes

B. *Variance explained by each component*

ξ_1:	2.42
ξ_2:	1.11
ξ_3:	.74
ξ_4:	.49
ξ_5:	.24
	5.00 Total variance explained

C. *Values of* ξ_1 *for each city*

City	ξ_1	City	ξ_1
Atlanta	−.23	Los Angeles	−1.21
Baltimore	.29	Milwaukee	−1.12
Boston	2.30	Minneapolis–St. Paul	−.45
Buffalo	−.35	New York	3.78
Chicago	.12	Philadelphia	.90
Cincinnati	.61	Pittsburgh	.62
Cleveland	−1.24	St. Louis	.23
Dallas	−1.12	San Diego	−1.93
Detroit	−.28	San Francisco	−.88
Honolulu	4.17	Seattle	−2.14
Houston	−1.32	Washington, D.C.	−.41
Kansas City	−.32		

would presumably be to use the first principal component as an index of how expensive food was in the different U.S. cities. Weighting bread, hamburger, milk, oranges, and tomatoes (and other foods if we used more variables) equally is really not apt to be appropriate. People buy more of some of these foods than they do of others. Surely it would make more sense to consider how much of various types of foods people actually consume and then to use the amounts consumed as importance weights. Doing so would give us a sensible index—the price of a standard market basket of goods. The procedure would not involve principal components at all.

To be even more exact in considering indices of the cost of food, we should note that it is likely unwise to consider data from one month of the year only. For one thing, buying patterns change over the year, as

TABLE 6–10

Based on Four Standardized Variables

A. *Definitions of the principal components*

$\xi_1 = .616 * \text{bread} + .635 * \text{hamburger} + .437 * \text{milk} + .164 * \text{oranges}$

$\xi_2 = -.121 * \text{bread} + .107 * \text{hamburger} - .333 * \text{milk} + .929 * \text{oranges}$

$\xi_3 = -.392 * \text{bread} - .266 * \text{hamburger} + .835 * \text{milk} + .279 * \text{oranges}$

$\xi_4 = .672 * \text{bread} - .717 * \text{hamburger} + .027 * \text{milk} + .180 * \text{oranges}$

B. *Variance explained by each component*

ξ_1:	1.95	
ξ_2:	1.02	
ξ_3:	.74	
ξ_4:	.30	
	4.01	Total variance explained

C. *Values of ξ_1 for each city*

Atlanta	.51	Los Angeles	−.45
Baltimore	.23	Milwaukee	−1.89
Boston	1.83	Minneapolis–St. Paul	−.78
Buffalo	−.70	New York	3.29
Chicago	−.03	Philadelphia	.06
Cincinnati	.94	Pittsburgh	.56
Cleveland	−1.43	St. Louis	.40
Dallas	−.86	San Diego	−1.11
Detroit	−.76	San Francisco	−.43
Honolulu	3.72	Seattle	−2.08
Houston	−1.04	Washington, D.C.	−.12
Kansas City	.12		

many customers are wise enough to buy nutritious foods that happen to be in season at a particular time. They are far less likely to buy fresh tomatoes in March than they are to buy fresh fruits and vegetables in August. Hence, we might want to construct an index of food costs by costing out a fully balanced nutritious diet that used whatever foods in specific categories were most available in a particular city at a particular time. Such an index would assign different consumption amounts to different particular foods at different times of year.

This process of designing ever more sophisticated indices could be continued. The main point about principal components analysis has been made, however. The construction of indices is often very useful. The application of principal components analysis to do so, however, weights each variable either by its variance (in the nonstandardized case) or equally (with standardized variables). Neither choice is likely to be satisfactory. In particular, with standardized variables, the relative importance of fruit prices and of meat prices in any resulting index will depend on just how many fruit variables and just how many meat variables have been used. For example, if tomatoes are not included, and only bread, hamburger, milk, and oranges are used, the first principal component with standardized variables is

$$\xi_1 = .616 * \text{bread} + .635 * \text{hamburger} + .437 * \text{milk} + .164 * \text{oranges}$$

(Table 6–10 provides additional results.) Now, the only fruit-or-vegetable variable, oranges, receives little weight. By comparison, in the five-food standardized variables case, tomatoes and oranges were used and tomatoes received the second-highest weight. It is generally far preferable to assign weights to variables in a more considered way in constructing indices. An example is to consider the amounts consumed, as has been sketched here for the food-prices example.

PRINCIPAL COMPONENTS AS UNCORRELATED VARIABLES

For completeness we should note that principal components analysis is sometimes used because it produces new variables which are uncorrelated with one another. In such cases, the technique is used for restatement rather than for data reduction.

Recall that in regression and other techniques, when the independent (or explanatory) variables are correlated with one another, we cannot unambiguously describe the effects of individual variables. In regression with correlated independent variables, we cannot unambiguously interpret the coefficients as the effects of their respective variables. Similarly, we cannot unambiguously decompose the overall R^2 of a regression model into contributions made by the individual independent variables.[10] With uncorrelated explanatory variables, on the other hand, the coefficients are the unique effects of their respective variables and the R^2 for the equation can usefully be broken down into the contributions of the individual variables.

Because of the desirability of uncorrelated independent variables, investigators sometimes manipulate an original set of variables to produce another set of restated but uncorrelated variables. Principal components analysis is most frequently used for this purpose. These investigators then use the principal components rather than the original variables in regression models and in other analyses.

The value of such an approach is questionable. It is certainly true that the coefficients and the contributions to R^2 of the principal components can be unambiguously defined. On the other hand, what the investigator has obtained is well-defined coefficients and contributions of variables which are not themselves likely to be easily understood. The procedure starts with meaningful (original) variables which produce hard-to-understand regression results. It then produces variables which can usually not be understood but which give unambiguous regression results. It is not at all clear what is the value of well-defined contributions to R^2 of variables which do not have clear intuitive meanings. It is, however, useful for readers to be aware that this use of principal components analysis is found in the literature.

---

[10]The reason is that with correlated variables both the coefficient and the marginal contribution to R^2 of a particular variable depend on which other variables are included in the equation.

CHAPTER 7

Factor Analysis

Factor analysis is a technique for analyzing the internal structure of a set of variables. The basic idea is that the members of a set of variables, each of which has been observed or measured for a number of observations, have some, though not all, of their structure determined by certain underlying, unobservable common constructs or *factors*. Investigators use factor analysis to help them identify such underlying constructs in sets of variables. One important characteristic of factor analysis is that it involves a set of variables each of which serves a similar role in the analysis. In particular, in contrast to many other multivariate techniques, factor analysis does not involve a variable singled out as the dependent variable.

The basic ideas of factor analysis were developed in the context of examination grades. This chapter will develop the ideas of the technique with such examples. Suppose that we have available the test grades of a large number of high school students for a set of examinations. Suppose that each student took one test in each of six subject areas: mathematics, chemistry, English, history, French, and Latin. The initial database is sketched in Figure 7–1, which shows six variables (the test grades) and many observations, one for each student who took the tests.

In introducing the ideas of factor analysis, Spearman in 1904 hypothesized that examination grades for an individual were related to one another.[1] He suggested that the link among a person's grades was a latent one, not directly measurable. It was, he believed, the individual's general level of intelligence. The test grades for one person were not completely correlated with one another; a person might, for example, be more able in English than in mathematics. On the other hand, Spear-

[1]C. Spearman, "General Intelligence Objectively Determined and Measured," *American Journal of Psychology*, 1904, pp. 201–93.

FIGURE 7–1

	Mathematics grade	Chemistry grade	English grade	History grade	French grade	Latin grade
Person 1						
Person 2						
Person 3						
⋮	⋮	⋮	⋮	⋮	⋮	⋮

man's hypothesis suggested that grades in all of the subject areas would depend to some degree on general intelligence, which could be considered a factor common to all of the tests.

THE MODEL WITH ONE FACTOR

In the simplest model of factor analysis, there is just one common factor tying the variables together. Whatever part of the individuals' grades on a particular test is *not* explained by the common factor is assumed specific to that particular test. In other words, the general or common factor is assumed to account for all of the correlations among tests. Whatever part of a particular test is not linked to the common factor is also not linked to any of the other tests.

Factor analysis assumes linear additive models. For the simplest case, with one common factor, the model states that the variable for the grade in a particular subject area is a weighted sum of the general factor and a subject-area-specific part. The weights in the sum are characteristics of the subject area, not of the individual students who took the tests. In other words, if the weights for chemistry grades are estimated to be .6 for the common factor and .4 for the chemistry-specific part, then the .6 and .4 weights are applicable for all students. Different students will, however, have different actual values for the common factor and for the chemistry-specific value. (Some students will be more intelligent than others and some will have more chemistry-specific skills than others.) Thus, in using factor analysis, investigators would assume that the *nature* of the common factor (general intelligence) and the weights are the same for all students but that students possess different numerical values for the common factor and subject-area-specific parts.

The basic model underlying the simplest version of factor analysis is given by the following equation:

$$Z_j = a_j f + m_j + d_j u_j$$

In this equation Z_j is the jth variable, or test grade. The test grade Z_j consists of three parts. The first part, $a_j f$, is the part involving the common factor f. f can be thought of as an unobserved variable; it will have different values for different observations (or people taking the tests). Without loss of generality the factor is assumed to have mean 0 and variance 1. In the equation for the jth grade the value of the common factor is multiplied by the constant (or weight) a_j, called the *loading* of the jth test on the common factor.

The second part of the equation is m_j, the mean score on the test. This part of the equation is a constant peculiar to the jth test and applies for all students; m_j merely serves to put the grades into the appropriate range for that test. It is common in factor analysis to work with standardized test scores, with mean 0 and variance 1; for standardized test scores, m_j would be 0.

The third part of the equation for the grade Z_j is the term $d_j u_j$ above. Here u_j is assumed to be another unobserved variable. Rather than being an underlying variable common to all the tests, however, u_j is specific to the jth test. Without loss of generality, u_j is assumed to have mean 0 and variance 1. d_j is a constant, specific to the jth test. Different individuals have different values for the test-specific, unobserved variable u_j. All individuals' test scores are computed with the same constant d_j for the jth test.

To summarize: the above equation says that scores for the jth test are the sum of three parts. Two parts are specific to that test: the test mean m_j and the unobserved specific variable u_j multiplied by the constant d_j. The remaining part of the equation is the loading a_j times the common factor f.

The equation above uses only one subscript, j, to identify the test being considered. To emphasize the fact that Z_j, f, and u_j take different values for different students, we can also introduce a second subscript, i, to identify the specific student:

$$Z_{ij} = a_j f_i + m_j + d_j u_{ij}$$

(Student i's grade on the jth test is a_j times student i's value f_i of the common factor plus the mean m_j of the jth test plus d_j times student i's test-j-specific value u_{ij}.) This text will use equations with fewer subscripts, but readers who find it useful can think in terms of adding an additional subscript to identify the student (or observation).

To understand the basic equation better, it is useful to think about what the equation implies for the process of determination of test scores in the real world. For this purpose, assume (unrealistically) for a moment that we can magically discover students' values for the common factor and subject-area-specific values. Suppose we consider one student, called A. Suppose that A's level of general intelligence, or common factor, is 1. Recalling that the factor values are treated as having mean 0 and variance 1 (also standard deviation 1), we note that student A has general intelligence exactly one standard deviation above the average. Suppose that we magically learn that A is somewhat weaker in the skills specific to test j and has a value of u_j equal to $-.5$. This value is ½ standard deviation below the mean or average score of subject-area-specific abilities. To find student A's actual grade on the jth test we would take the constant a_j times 1, plus the test mean m_j, plus the constant d_j times $-.5$. For a second student, B, we would use an analogous procedure but would substitute B's individual values for the general factor f and the test-j-specific ability u_j to obtain the final grade on test j.

In practice, of course, we cannot observe either the values of the factor (the f values) or the values of the test-specific-ability variables. Instead, we observe only the actual test grades, the Z_js. Among the purposes of factor analysis is the exploration of how tests are related by a common factor. In other words, we want to use our information about the observed variables (the actual test grades) to infer information about the unobserved common factor and the test-specific unobservable variables. Factor analysis provides estimates of the loadings (a_js) and also of the unobservable values of f for individual students. These f values are called *factor scores*.

Before proceeding to consider more complex factor analytic models, there are a few additional points that can be made with this especially simple model. First, it is worth repeating the fact that the tests are assumed to be linked *only* through the common factor. In particular, the unobserved test-specific variable u_j for the jth test and the unobserved test-specific variable u_k for the kth test must be uncorrelated with one another. The common factor is assumed to account for *all* correlation among test grades.

On the other hand, the model does allow the common factor to be more important in determining the grade on one test than it is in determining the grade on another test. (The loading a_j may be very different from the loading a_k.) For example, for the six high school tests in the current example, it may be that students' grades on the mathematics test are very substantially explained by their general level of intelligence, the common factor. Thus, a_1 would be relatively large. On the other hand, it may be true that general intelligence is less important in determining grades on the French test. While intelligence may explain some part of the French grades, it may be that French involves considerable listening and pronouncing skills which are not the same as the general intelligence factor. Thus, a_5 would be relatively small. In the terminology of factor analysis, that portion of the variability of individuals' grades on a particular test that is explained by the common factor is called the *communality* of that test. The hypothesis just advanced for the mathematics test as compared with the French test would be expressed as follows: The mathematics test has a higher communality than does the French test. Equivalently, the specific portion of the French-grade equation accounts for a larger fraction of the variability of French grades than the fraction of the variability of mathematics grades accounted for by the test-specific portion of the mathematics grades.

The basic model makes the assumption that the unobservable common factor f and the unobservable test-specific variable for any test are uncorrelated with one another. This fact makes it possible to express the total variance of the grades for a particular test as the sum of two parts. One is the communality of the test. The other is the variance of the test-specific portion of the equation for grades. In symbols, if $V(Z_j)$ is the variance of the grades on test j, then[2]

$$V(Z_j) = a_j^2 + d_j^2$$

It is also usual in factor analysis to consider the total communality for all tests. For the example with six tests, the total communality is

$$a_1^2 + a_2^2 + a_3^2 + a_4^2 + a_5^2 + a_6^2$$

This quantity must be less than the total of the variances of the six tests. (With six standardized test-grade variables, each with variance 1, that total variance is 6.)

It also turns out, because two different tests are linked only through the common factor, that the correlation between grades on two tests can be expressed in terms of the loadings of those tests on the common factor. In particular, the covariance of the jth and k^{th} tests is equal to $a_j a_k$. (When the test grades have been standardized, this quantity is also the correlation between the tests. For unstandardized grades, it is the correlation times the product of the standard deviations of the two tests.)

In essence, this simple factor analytic model assumes that the correlations among a set of test-grade variables are explained through a common factor f. It should not be surprising, therefore, that the actual calculations in factor analysis operate on the observed correlations among test grades. The basic aims of those calculations for the case of one common factor are:

---

[2]In determining this relationship, we also use the fact that f and u_j are taken to have standard deviations of 1.

1. To determine the degree to which each test depends on the common factor. In other words, to estimate the loading of each test on the common factor.
2. To estimate the values of that common factor for individual students. In other words, to estimate students' general intelligence from their grades on the individual tests. These factor values are called factor scores.

MODELS WITH MORE THAN ONE FACTOR

Before proceeding to consider how programs for factor analysis achieve these basic aims for the one-factor model, we turn now to a consideration of factor analytic models with more than one factor. In the example of test grades on six high school subjects, we may well feel that one factor, general intelligence, does not adequately explain the relationships among the grades. Knowing people who are much better at mathematics and sciences than they are in the humanities and other people who perform considerably better in subjects in the humanities than they do in the sciences, we might instead believe that there are two underlying types of ability. We might expect to find separate factors for quantitative and for verbal abilities.

To understand the structure of a factor analytic model with two underlying factors, we can consider what such a model might imply for the underlying process by which test grades are generated. (Again, however, it is important to realize that this discussion is for motivational purposes only. We cannot actually observe factor scores, and the normal procedure in factor analysis is to start with observed test grades and to try to learn from those grades about the underlying factors.)

With two underlying factors, the basic equation for the grade Z_j on the jth test is the following:

$$Z_j = a_{j1}f_1 + a_{j2}f_2 + m_j + d_ju_j$$

Here m_j is the mean for the jth test, just as it was in the single-factor model. With standardized test grades, m_j would be 0. Similarly, u_j and d_j retain essentially the same meanings that they had in the single-factor model. u_j is an unobservable variable that measures abilities specific to the jth test. The specific variable u_j for the jth test is uncorrelated with the specific variable for any other test, such as the kth test (u_k). In addition, u_j is uncorrelated with the common factors (the fs). Usually, each unobservable specific variable is assumed to have mean 0 and variance 1. d_j is then a constant by which the specific variable for the jth test should be multiplied.

The only real difference between this two-factor equation and the equation for the one-factor model is that we now have two common factors, f_1 and f_2. Generally, each of these factors is assumed to have mean 0 and variance 1. Often, the two factors are assumed to be uncorrelated with one another. (The following discussion will return to consider briefly the possibility of correlated common factors but for now we will assume that the factors are uncorrelated.) a_{j1} is called the loading of the jth test on the first common factor and a_{j2} is the loading of the jth test on the second common factor.

Suppose that the first factor measures quantitative ability and that the second measures verbal ability. The model assumes that an individual student's grade on some test, such as the English test, is determined from that student's values for the common factors and the test-specific part. While these values are assumed unobservable, to understand the model it is useful to pretend (unrealistically) that we know the student's values on the quantitative factor f_1, the verbal factor f_2, and the English-specific variable u_3.[3] We would then add the mean English grade m_3, plus d_3 times the student's English-specific value u_3, plus a_{31} times the quantitative factor f_1, plus a_{32} times the verbal factor f_2 for the student. We would expect that the English test grade would have relatively little relationship to a student's quantitative abilities; in other words, we would assume that a_{31} would be very small. On the other hand, if it turned out that the ability to understand logical arguments was of some use in English, and that such ability was part of the quantitative factor, we might find some small loading for the English test on the quantitative factor. Similarly, while we would expect the mathematics test to load more heavily on quantitative ability than on verbal ability, it might nevertheless be useful in mathematics to be able to read and write well, so the mathematics test might show some small loading on the verbal factor.

When the factors in a two or more factor model are assumed uncorrelated with one another, the variances of the individual test variables can be decomposed into parts. If $V(Z_j)$ is the variance of students' grades on the jth test, then[4]

$$V(Z_j) = a_{j1}^2 + a_{j2}^2 + d_j^2$$

The variance or those test grades is the sum of three parts. The first term, a_{j1}^2, is called the communality of the jth test due to (or contributed by) the first factor. The second term, a_{j2}^2, is the communality of the jth test contributed by the second factor. The sum of these two terms is called the *communality of the* j

---

[3]If we added a subscript i to denote the values for the ith student, we would label these values f_{i1}, f_{i2}, and u_{i3}.

[4]This formula involves the assumption that each of the factors and the u_js has variance 1.

The *total communality* for the two-factor model is defined as the sum of the communalities of the individual variables. In the current example, the total communality is the sum of the communalities of each of the six tests. It is also usual to consider the portions of the total communality that are contributed by (or due to) each of the factors. In the current example, the first factor contributes $a_{11}{}^2$ to the communality of the first variable (or test), $a_{21}{}^2$ to the communality of the second variable, $a_{31}{}^2$ to the communality of the third, and so on. Hence,

$$a_{11}{}^2 + a_{21}{}^2 + a_{31}{}^2 + a_{41}{}^2 + a_{51}{}^2 + a_{61}{}^2$$

is the total contribution of the first factor, or the communality due to that factor. Similarly,

$$a_{12}{}^2 + a_{22}{}^2 + a_{32}{}^2 + a_{42}{}^2 + a_{52}{}^2 + a_{62}{}^2$$

is the communality contributed by the second factor.

As was true for the one-factor model, the correlations among the test grades for standardized test variables (or the covariances for unstandardized ones) can be expressed in terms of the loadings of the tests on the factors. This fact results from the assumption that the factors account for all correlation among the test variables. In particular, the covariance between the jth test and the kth test is equal to

$$a_{j1}a_{k1} + a_{j2}a_{k2}$$

(For standardized variables this expression is also the correlation between the two test variables.)

This two-factor model can easily be expanded to a model with more than two factors. For example, the following equation gives the expression for the grade on the jth test with a three-factor model.

$$Z_j = a_{j1}f_1 + a_{j2}f_2 + a_{j3}f_3 + m_j + d_ju_j$$

Again, the unobserved variable u_j is specific to the test. The factors account for all of the correlations among the test grades. u_j is uncorrelated with u_k for any other test. u_j is also uncorrelated with each of the three factors. f_1, f_2, f_3, and the *u*s are all assumed to have mean 0 and variance 1. Finally, for the time being we continue to assume that the factors f_1, f_2, and f_3 are uncorrelated with one another. Then, the communality of the jth test is equal to

$$a_{j1}{}^2 + a_{j2}{}^2 + a_{j3}{}^2$$

The variance of that test is equal to the communality plus $d_j{}^2$. The total communality is the sum of the communalities of the individual tests. This total is often expressed as the sum of the communalities due to each of the three factors. The covariance between the jth test and the kth test is

$$a_{j1}a_{k1} + a_{j2}a_{k2} + a_{j3}a_{k3}$$

The factor-analytic definition can also be extended to models with more than three factors.

AIMS OF FACTOR ANALYSIS

We return now to the usual situation in factor analysis in which we have observed values for a set of variables for each of a group of individuals. We do not, however, know what factors link those observed variables and, in fact, we do not even know how many factors there are. Further, we cannot observe the scores of individuals on the underlying factors.

As an example, we will use a database consisting of information on the final grades of a class of MBA students in 10 courses. Table 7–1 lists the 10 final-grade variables along with a brief description of each course. The database contains 753 usable observations (that is, there is a complete set of grades for each of 753 students).

The basic aims of factor analysis are:

1. To determine how many factors underlie the set of variables.

2. To find the extent to which each variable (final grade) depends on each common factor. In other words, we want to find the loadings of the grades on the factors. The loading of the jth grade on the pth factor was called a_{jp} in the equations above for the factor analytic models.

3. To interpret the factors.

TABLE 7–1
List of Variables (courses)

Variable	Description
1. Finance Final	Final grade in Finance course
2. HBO Final	Final grade in Human Behavior in Organizations (organizational behavior course with emphasis on the individual)
3. OP Final	Final grade in Organizational Problems (organizational behavior course with emphasis on organizational structure)
4. LOB Final	Final grade in Laboratory in Organizational Behavior
5. MERC1 Final	Final grade in Control (accounting)
6. MERC2 Final	Final grade in Managerial Economics (quantitative methods)
7. MKTG Final	Final grade in Marketing
8. PBE Final	Final grade in Planning in the Business Environment (macro economics, government relations, and related topics)
9. PROD Final	Final grade in Production and Operations Management
10. WAC Final	Final grade in Written Analysis of Cases (written compositions)

4. To find the amount of each common factor possessed by each observation: the factor scores. (In the final-grades example, this aim becomes that of finding the amount of each underlying factor possessed by each student in the class.)

CHOOSING THE NUMBER OF FACTORS

In using factor analysis the investigator must, in one way or another, specify the number of factors to be considered. Since we normally begin an analysis without knowing how many factors to expect, this requirement is not an easy one. The following principle should guide the choice. It is important for the investigator not to leave out any important factors. In other words, if there are really five important underlying factors and if the investigator instructs the program to consider four factors, all of the results will be basically worthless. The program cannot find the loadings for four of the five true factors. On the other hand, if there are five important factors and the investigator tells the program to consider six, the program can find information about the five true factors plus information about a sixth unimportant or even spurious factor. That factor will appear on the program output to contribute little to the explanatory power of the factor analysis model; in other words, it will account for little of the total communality of the model. Accordingly, the investigator can drop it from further consideration.

Even if investigators want to consider only one or two of the common factors in a model, they must perform the calculations to include all factors that make important contributions to the communality of the model. Thus, the general principle is to be sure to include as many factors as are likely to contribute significantly to the communality. Users are generally not comfortable with this rather imprecise instruction to include enough factors. Fortunately, many factor analysis programs provide specific output that helps indicate whether or not enough factors were considered. The preceding discussion stated that factor analysis programs require as input the correlations among the variables. It also noted that once a factor analysis model has been fit, the values found for the loadings can be used to calculate implied correlations among the variables. Many programs provide output on both the input correlations and the implied correlations. It turns out that if the two sets of correlations are almost the same, the investigator can feel comfortable about having used enough factors. If, however, the two sets turn out not to be close, the investigator should try using more factors.

An alternative method for specifying the number of factors that is sometimes available is to tell the program to find successive factors until no more factors can be found that account for at least a user-specified portion of the communality. This procedure is a direct statement of the requirement that the only factors that can sensibly be omitted from consideration are those that do not contribute heavily to the communality.

The discussion in this section likely leaves the reader feeling a bit lost as to how to choose the number of factors for consideration. The feeling is appropriate, for there are really no hard rules. The best procedure may be to try several numbers of factors and to study model output carefully, in ways described above and below.

The problem is complicated by the fact that for any particular number of observed variables there is an upper limit to the number of factors that can be considered. The formulas for those limits will not be considered here, but the reader should be aware that a factor analysis program may be unable to consider a large number of factors for a particular set of data.

FINDING THE CANONICAL FACTORS

The specific choices available to the user for specifying the number of factors depend on which of several calculation procedures are being used to find a set of factor loadings. Hence, we now turn to a consideration of such procedures and then continue the discussion of the number of factors in that context.

The calculations of factor analysis are complicated by the fact that in general there are different ways of describing a set of common factors. (The issue will be pursued below. For now, it is enough to note that the problem is essentially that of choosing a set of coordinate axes for describing the locations of a group of points. Many sets of axes will do.)

Most factor analysis programs start by determining what are called *canonical factors*. Such factors are uncorrelated. They are defined so that the first factor accounts for the largest possible communality, the second factor the largest possible part of what remains, and so on. Finally, canonical factors are defined so that the sum of the positive loadings for any one factor exceeds minus the sum of the negative loadings for that factor. The important point here for users of factor analysis is that programs must select one of many possible ways of representing a set of a specified number of underlying factors. The usual choice is to start with uncorrelated factors that account for successively smaller amounts of the total communality of the variables.

One very good procedure for calculating the canonical factors in-

*TABLE 7–2**

Grade	Grade 1	2	3	4	5	6	7	8	9	10
1	1.0000									
2	.2041	1.000								
3	.2977	.4223	1.000							
4	.2297	.1496	.1940	1.000						
5	.5168	.2535	.2665	.1606	1.000					
6	.5106	.1867	.3043	.1448	.5993	1.000				
7	.4687	.3715	.3310	.2296	.4674	.4438	1.000			
8	.3325	.4093	.3523	.1344	.2844	.2866	.3539	1.000		
9	.4750	.2591	.3048	.2749	.4478	.4750	.4926	.3386	1.000	
10	.3548	.1960	.2823	.1912	.3303	.3156	.3371	.2867	.3617	1.000

* The correlation between the jth grade and the kth grade is, of course, equal to that between the kth grade and the jth. Therefore, only half the off-diagonal entries in a correlation matrix such as this need be filled in.

volves what is called the *minimum residual method.* The procedure starts with the correlation matrix of the observed variables.[5] For the final-grade example, it starts with the correlation matrix shown in Table 7–2.

The procedure must be told how many factors to consider. (The user gives a specific number.) The program then conducts a search procedure to determine the loadings (the a_{jp} values) and the d_js for the model. Its aim in searching is to find a model that agrees as closely as possible with the observed correlations in the input correlation matrix. Recall that in the discussion of the one-factor and two-factor models we noted that the a_{jp} and d_j coefficients could be used to specify the implied correlations and communalities of the variables of the model. Thus, the minimum residual procedure of trying to find loadings and d_js which give implied correlations that match the observed (or input) correlations should seem reasonable.

A second method for finding factor loadings is called the *principal factor method.* This method is very frequently used, likely because it involves considerably less computation than do alternative procedures. The computational advantage was important before computers were available to perform factor analyses. It is much less important today. Moreover, the principal factor method has the distinct disadvantage that the user must somehow specify the communality of each of the variables. In other words, the user must give a rule for deciding how

[5]Most multivariate techniques instead begin directly with lists of observations, and discussions of those techniques include warnings that there should be enough observations. In factor analysis, the appropriate warning is that there must be enough observations used in finding the correlation matrix. If, for example, we had calculated correlations using a small number of students, the values would likely be quite different from those found with the full set of students.

much of the variance of each variable is assumed to be explained by the common factors.

With the principal factor method the user can either give a number specifying how many factors should be considered or, alternatively, can specify that the program should find successive factors until it cannot find any more factors that account for at least a specified fraction of the total communality. (This second choice is not available with the minimum residual method.)

As for the specification of the communalities, users will very rarely indeed be able to specify the numbers directly. Several choices are usually possible. One is to assume that all communalities are 1 (in other words, that the common factors account for *all* of the variances of the variables). Another is to assume that the communality of any particular variable should be obtained as the R^2 from a regression of that variable on all of the other variables. For reasons that are not intuitive and are not explained here, the second choice is a good one when there are many more variables than factors and when some of the communalities are high. A very common third choice is to take the average of the values for communalities from the first two methods.

Again, the reader is likely feeling very uncomfortable with this discussion. Again, the feeling is appropriate. Moreover, the problem here is a substantial one. If the user specifies communalities that are far from the true communalities among the variables, the program results will be worthless. Users of the principal factor method must be sure to check the agreement between their assumed communality values and the final values implied by the models. (This step is often omitted when the principal factor method is used. As a result, model results are often highly suspect.)[6] More basically, however, this discussion argues strongly for the use of the minimum residual method wherever possible.[7]

---

[6]Investigators who for some reason must use a principal factor procedure can sensibly use such a program iteratively until the specified and implied communalities are very close. In other words, investigators can start with a set of specified communalities—perhaps even all 1s. They can then run the program and observe the implied (output) communalities. If, as is likely, the input and implied communalities are not close, the users can give the implied communalities from the first run as input communalities for the second factor analysis. If the input and implied communalities on the second run are not sufficiently close, the output communalities from the second run can be the inputs for the third, and so on. The iterations would stop when (or, more accurately, if) the two sets of communalities on a single run agreed closely.

Such iterative use of the principal factor method unfortunately does not seem to be common. Instead, published reports of the use of this technique seem to involve only a single run—and seem to omit the important step of comparing input and implied communalities.

[7]There are other methods, too. One is the *maximum likelihood method,* which involves an assumption of normality of the factors and the u_js. This method is not considered in this chapter.

CANONICAL FACTORS FOR THE FINAL-GRADES DATA

Having immersed the reader to a considerable depth in the messy details of the calculation procedures for factor analysis, we can now at last turn to consider some results from a factor analysis program. Table 7–3 shows the loadings on the canonical factors when the MBA final-grade data were subjected to a factor analysis involving three factors with the minimum residual method. For example, the first row of that table should be read as follows:

> The finance final grade is estimated to be .679 times a first factor minus .202 times a second plus .056 times a third. (Plus d_1 times a finance-specific value. d_1 is not shown in this table.)

Table 7–4 shows the communality for each variable. We can find the d_j^2 values from this table. For example, because we know that the

TABLE 7–3

	Loadings On		
Final Grade	*Factor 1*	*Factor 2*	*Factor 3*
1 FINANCE	.679	−.202	.056
2 HBO	.517	.637	−.166
3 OP	.516	.258	.040
4 LOB	.316	.027	.261
5 MERC1	.697	−.254	−.241
6 MERC2	.686	−.296	−.190
7 MKTG	.683	.015	.044
8 PBE	.527	.231	.023
9 PROD	.678	−.105	.183
10 WAC	.502	−.029	.153

TABLE 7–4

Variable	*Communality*
1 FINANCE	.50
2 HBO	.70
3 OP	.33
4 LOB	.17
5 MERC1	.61
6 MERC2	.59
7 MKTG	.47
8 PBE	.33
9 PROD	.50
10 WAC	.28

TABLE 7–5

Factor	*Contribution (variance explained)*	*Relative Contribution (fraction of total communality)*
1	3.506	.780
2	0.731	.163
3	0.255	.057
		1.000

variance of the standardized HBO score is 1 and that this variance is the communality (.70) plus d_2^2, we now know that d_2^2 is estimated as $1.0 - .7 = .3$. In other words, .3 of the variability in HBO grades is estimated to be specific to HBO.

Table 7–5 shows how much of the total communality (over all 10 grades) is accounted for by each of the factors. Table 7–6 shows the differences between the input correlations (from the data) and the correlations among the grades that are implied by the factor loadings found by the program. (The minimum residual calculations were aimed at reducing these residuals to be as small as possible.)

The most interesting use of these program outputs is generally in the interpretation of the factors. Often, investigators do not try to interpret the canonical factors but instead instruct the computer to prepare alternative descriptions of the results for interpretation. (Such alternative descriptions are discussed in the following section.) For the current example, however, we will consider possible interpretations of both the canonical representation and an alternative one.

The job of interpretation must be done by the investigator rather than by the program. The procedure is to examine the loadings, think about the variables, and try to interpret the results. In the process, in-

TABLE 7–6
Residuals (input correlation minus model correlation)

	1 FINANCE	*2 HBO*	*3 OP*	*4 LOB*	*5 MERC1*	*6 MERC2*	*7 MKTG*	*8 PBE*	*9 PROD*	*10 WAC*
1 FINANCE	.0000									
2 HBO	−.0091	.0000								
3 OP	−.0026	−.0023	.0000							
4 LOB	.0060	.0123	.0135	.0000						
5 MERC1	.0058	.0144	−.0179	.0103	.0000					
6 MERC2	−.0044	−.0118	.0342	−.0143	−.0000	.0000				
7 MKTG	.0058	.0164	−.0267	.0019	.0061	−.0118	.0000			
8 PBE	.0204	−.0064	.0201	−.0443	−.0183	−.0020	−.0099	.0000		
9 PROD	−.0168	.0059	−.0251	.0158	−.0073	.0135	.0233	.0018	.0000	
10 WAC	−.0004	−.0201	.0246	−.0067	.0101	−.0082	−.0120	.0254	−.0096	.0000

vestigators must use their own knowledge of the variables being considered. For a particular factor, they observe the pattern of loadings, noting which are large or small and which are positive or negative. They then try to intuit a meaning or interpretation for the factor that would correspond to such a set of loadings.

For the output in Table 7–3 we note that all of the grades load positively on the first factor (although some load more heavily than others). We might conjecture that this factor is measuring general academic ability. We might then note that HBO loads particularly heavily in a positive direction on the second factor, while OP and PBE also have substantial positive loadings. FINANCE, MERC1, and MERC2 show heavy negative loadings. This factor might be measuring ability in the courses that emphasize behavior and people more than they do numbers.

The loadings on the third factor are harder to interpret. HBO, MERC1, and MERC2 show sizable negative loadings, while LOB, PROD, and WAC show relatively large positive ones. The nature of the factor is not clear. On the one hand, we might argue that the third factor does not contribute heavily to the total communality and might safely be ignored. On the other hand, it would seem prudent to pursue the analysis further before taking such a step.

Part of the problem in interpreting this third factor may be that it shows large loadings on so many variables. Often in factor analysis it is useful to consider some of the alternative expressions for the information in a set of factors. In doing so, we hope to find a more easily interpreted set. The next section develops this idea of alternative or derived factors and then applies the idea to the current example.

DERIVED FACTORS

The preceding discussion noted that there are many ways of expressing the results of a factor analysis of a set of data. While programs often start with the representation in terms of canonical factors, it is not at all clear that that particular representation will be best suited to help the investigator in a specific problem interpret the factors. Consequently, it is common in factor analysis to proceed past the first set of canonical factors and to consider alternative representations of the results in an attempt to improve interpretability. The alternative representations obtained in this process are called *derived factors*.

Some readers will find it useful motivation for what is to follow to consider the procedures for finding derived factors that were used by early investigators (before computers). Suppose that we have results from factor analysis calculations for six tests and two factors. We can construct a graph, plotting one point for each test. The point is located

FIGURE 7–2

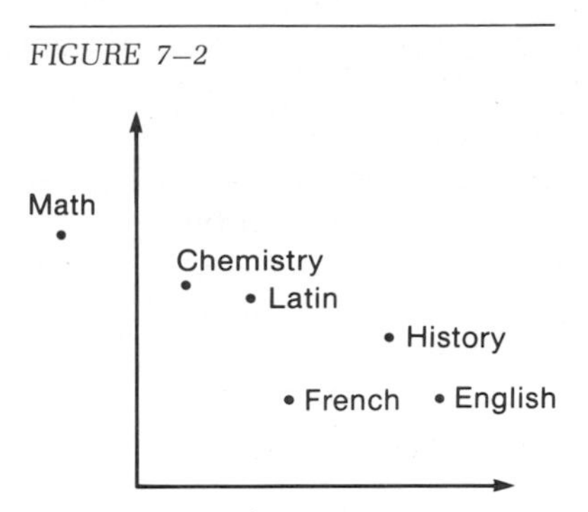

with its coordinates equal to the loadings of the test on the two canonical factors. In other words, the calculations have found values for the two loadings a_{11} and a_{12} for the first test. We plot a point at a_{11} on the horizontal axis and a_{12} on the vertical axis to represent the test. Similarly, we plot a point for each other test. Figure 7–2 shows a hypothetical set of results.[8]

It turns out that the information in these results could equally well be represented relative to another set of axes. Moreover, some other set of axes may give factor loadings (or coordinates of the points in Figure 7–2) that are more easily interpreted. Early investigators would look at figures such as Figure 7–2 and would try to choose new axes that were more easily interpreted. For example, Figure 7–3 shows a possible alternative set of axes for the points in Figure 7–2.[9] The almost vertical axis might now be interpreted as quantitative ability and the other axis as verbal ability. For example, the math test has high loading on the quantitative axis and little loading in the verbal direction. The reverse is true for the English test.

Early investigators had no alternative but to consider such plots and changes of axes. Even when they used more than two factors, they

[8]It turns out that the communality of any test is equal to the squared length of the line from the origin in Figure 7–2 to the point for that test.

[9]It is worth noting that the points could also be given relative to axes that were not perpendicular. Such a procedure is certainly rare in graph-drawing in general, but it may aid interpretability in factor analysis. Choosing nonperpendicular axes is equivalent here to assuming that the two factors may be correlated with one another. (Such factors are called *oblique*.) There is considerable reason to think that factors may be correlated in practice. In the six tests above, it would indeed seem likely that the two abilities were correlated. (Individuals with high quantitative ability often, though not always, also have good verbal ability—and vice versa.) In practice, however, uncorrelated factors are much, much more frequently considered—in large part because they can be found in more mechanical ways by computer than correlated factors can be. Unfortunately, investigators seem too frequently today to make choices (such as the choice of perpendicular axes) that are convenient for computerization without worrying about whether or not the choices are sufficiently realistic for the particular problems under analysis.

FIGURE 7–3

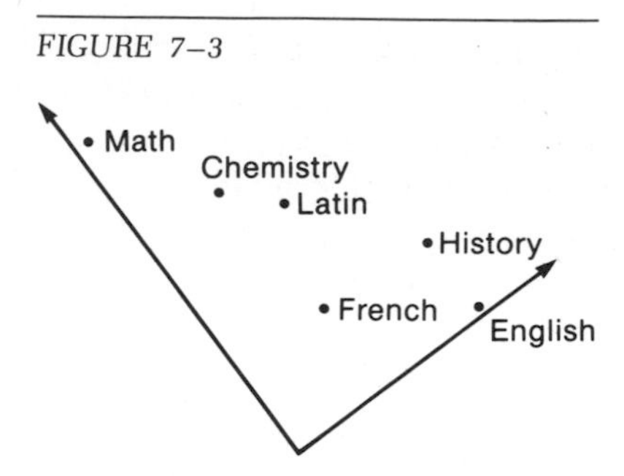

considered plots of two at a time. With the widespread availability of computer programs for factor analysis, the procedure of finding alternative sets of factors has been mechanized. The procedures to be described below are simply attempts (in fact, rather awkward ones) to make routine the procedure an analyst would go through in trying to select new axes in graphs such as Figure 7–2.

Basic to the procedures for finding derived factors is what is called *simplicity*. It is often considered helpful in interpreting factors if the loadings show the following properties:

1. For each variable, there are some factors on which that variable does not load heavily. In other words, in each row of the matrix of factor loadings there are some values near zero.
2. With the possible exception of one general factor (such as general academic ability, above) each factor shows low loadings for some variables. In other words, except perhaps for a general factor, there is no factor on which all variables load heavily.
3. With the possible exception of one general factor, different factors have large loadings for different sets of variables.

Attempts to find derived factors which exhibit some form of such simplicity are not easy. The basic problem is in giving the computer program a completely specific set of instructions for selecting an alternative representation of the factor loadings. Generally, investigators proceed to consider derived factors without considering whether a given set of data should in fact produce a simple structure.[10] In addi-

[10]Some investigators would argue that one should always consider derived rather than canonical factors. Largely because (1) the assumption of simplicity is too often made implicitly rather than explicitly and (2) the particular definitions of simplicity used in computer programs are not very intuitive, this chapter instead includes interpretations of both canonical and derived factors. The idea is to decrease, at least a bit, the emphasis on *mechanically* derived factors.

tion, investigators very often assume uncorrelated factors without explicitly considering whether such an assumption is reasonable.[11]

When a user of factor analysis does in fact believe that it is appropriate to try to find a set of uncorrelated derived factors exhibiting more simple structure, there are two common criteria of simplicity that can be made operational on a computer. Each involves consideration of the variance of the individual factor loadings for a set of factors. Since the variance of a set of loadings will be larger for a set involving both small (near zero) and large loadings, criteria that maximize various variances of loadings are considered to help produce simplicity.

The first of the two criteria for simplicity is called the *varimax criterion*. Suppose that we have fit an initial model with m factors to a set of n tests. The factor loadings will form a matrix with n rows (one for each test) and m columns (one for each factor). Figure 7–4 sketches the matrix.

For the varimax criterion we find the variance in each of the columns of this matrix. Thus, we find m variances. The program is instructed to find the set of uncorrelated factors which gives the highest total (sum) of these m variances. This criterion encourages both large and small loadings in the column for each factor. Thus, it is particularly well suited to producing the second of the three aspects of simplicity listed above. At the same time, users should be aware that this criterion will tend to suppress the appearance of any general factor because a general factor, by definition, tends to have high loadings on all variables. Thus, the varimax criterion should not be used when the user suspects the presence of a general factor.[12]

When it finds a new set of uncorrelated factors through the varimax (or other) method, the factor analysis program will produce a new set of loadings, which the investigator can then try to interpret. The program will also provide new values for the communalities contributed by each factor. The program will not, however, produce changed values for the communalities of the variables, because those values are not changed by the rotations in finding new sets of uncorrelated factors.

The second criterion for defining simple structure operationally within computer programs for factor analysis is called the *quartimax criterion*. In the loadings matrix in Figure 7–4 we calculate the variance of the loadings across each row (not down each column as we did for the varimax criterion) and then sum those variances. Under the

---

[11] As noted in footnote 8, it is not at all obvious in the final-grade example that the factors underlying performance are not correlated with one another—in other words, that an individual who is high on one factor would not also likely be high on other factors.

[12] In fact, we can make a stronger point. The varimax criterion tends to produce factors with similar average squared loadings.

FIGURE 7–4

Factor Loadings

	factor 1	factor 2	. . .	factor m
variable 1	a_{11}	a_{12}	. . .	a_{1m}
variable 2	a_{21}	a_{22}	. . .	a_{2m}
.	.	.	. . .	.
.	.	.	. . .	.
.	.	.	. . .	.
variable n	a_{n1}	a_{n2}	. . .	a_{nm}

quartimax criterion, we instruct the computer to find the set of uncorrelated factors which gives the highest sum of these n variances. The effect of the quartimax criterion is to increase the variability of the loadings throughout the matrix.[13] The criterion does not suppress the appearance of a general factor and hence it can be used when the investigator suspects the presence of such a factor.

Table 7–7 shows the loadings found with the quartimax criterion for the final-grades data. It also gives the communalities contributed by each of the new derived factors.

The first factor in Table 7–7 still seems to measure overall academic ability. The second factor now shows large loadings for HBO, OP, PBE, and MKTG (and perhaps WAC and PROD). It seems to measure ability in the people-oriented courses. Unfortunately, the third factor does not seem more interpretable than did the earlier (canonical) third factor. It shows relatively large positive loadings on LOB, PROD, and WAC, somewhat smaller ones on OP, MKTG, and PBE, and relatively large negatives on MERC1 and MERC2. For this particular problem, use of the quartimax criterion has not appreciably improved interpretability.

The varimax criterion would not be appropriate for use with this example if we believed in the presence of a general factor (overall academic ability). Otherwise, we could try that criterion as well.

[13]It may seem strange that the criterion affects the variability of loadings throughout the matrix rather than within each row. The reason is that the variability within a row is closely related to the communality of that variable, and the communality cannot be changed by such a rotation.

TABLE 7–7

Grade	Factor 1	Factor 2	Factor 3
1 FINANCE	.704	.032	.092
2 HBO	.269	.793	−.023
3 OP	.391	.406	.126
4 LOB	.276	.088	.292
5 MERC1	.751	.040	−.207
6 MERC2	.753	−.010	−.164
7 MKTG	.632	.236	.112
8 PBE	.412	.388	.107
9 PROD	.664	.100	.230
10 WAC	.474	.117	.195
Communality contributed by this factor	3.164	1.034	.295
Relative communality contributed by this factor	.704	.230	.066

TABLE 7–8*

	Student							
	1	2	3	4	5	6	7	8
1 FINANCE	69	84	84	66	75	75	78	81
2 HBO	75	72	90	78	72	75	78	81
3 OP	75	78	81	75	66	87	81	87
4 LOB	84	84	87	75	78	81	75	81
5 MERC1	66	75	84	75	84	81	84	87
6 MERC2	78	72	84	72	81	81	84	78
7 MKTG	66	75	87	72	78	81	81	84
8 PBE	81	75	84	78	66	75	78	90
9 PROD	75	78	87	60	69	84	81	87
10 WAC	75	75	87	72	78	75	78	90
Factor 1	−.884	.185	2.040	1.344	.365	.962	1.219	1.882
Factor 2	.060	−.587	1.605	.616	−1.177	−.122	.052	.844
Factor 3	.335	.716	.612	−1.460	−1.494	.037	−.812	.749

*Grades are shown without standardization.

FACTOR SCORES FOR INDIVIDUALS

Factor analysis programs can also estimate the values of a set of factors (canonical or derived) possessed by each individual in the original database. (These values are estimates of the individuals' factor scores.) Table 7–8 shows results for 8 students from the 753 in the original sample. For each of these students, the table lists the grades in each course (which were included in the input data to the program), together with the program's estimates of the student's scores on each of the factors. The factors used are the ones described in Table 7–7 (those found with the quartimax criterion).

Factor scores determined by programs for factor analysis can be used by investigators as variables in additional multivariate analyses—with cross-tabs or regression, for example.

SUMMARY

In closing this discussion it is useful to recall where we started. Factor analysis is a technique for analyzing the *internal* structure of a set of variables. The technique is unfortunately also misused for other purposes as well. In particular, it is sometimes misapplied by investigators who want to group variables which are in some overall sense similar to one another. As described here, factor analysis is a technique that helps in identifying underlying unobservable dimensions *within* variables. It can be thought of as a technique for dissecting or looking within observable variables. When investigators instead want to group variables that are similar, they should use cluster analysis rather than factor analysis. Because of the common misuse of factor analysis in place of cluster analysis, this book includes a more extensive comparison of the two techniques; the comparison is presented in the appendix at the end of the chapter on cluster analysis.

CHAPTER 8

Cluster Analysis

Cluster analysis involves the grouping of entities that are similar to one another. The entities may be observations. For example, we might collect information on the consumption habits of individuals, with one observation for each individual. We could use variables for the dollar amounts spent on various categories of goods and services, such as clothing, food, vacations, and so on. We might then cluster the individuals into groups whose consumption patterns were similar. In other cases, the entities on which we perform cluster analysis to group similar entities may be variables rather than observations. For example, in the early stages of a market research project we might have a large number of questions that we are considering asking of a large sample of respondents. To reduce the number of questions needed, we could as a first step administer a long questionnaire, with all of the questions, to a smaller sample of people. We could construct a database with one observation per person in the initial sample and one variable corresponding to each question in the full questionnaire. We could then cluster the questions (or variables) to find which ones were really redundant, in that their answers were very highly correlated with the answers to other questions.[1] This information could then be used to weed out questions from that list.

Several decisions must be made by users of cluster analysis. For one thing, they must first decide that it is worthwhile to use formal clustering rather than simply to group their observations or variables on the basis of inspection and judgment. Then, the users must decide how the similarity or dissimilarity between entities is to be measured. They must also select one of many possible procedures for constructing

[1]For this procedure to make sense the responses (variables) would have to be of types for which calculating correlations made sense. Cardinal scales would be appropriate.

clusters of entities. In fact, cluster analysis might best be thought of as a collection of techniques for grouping entities.[2] One of the jobs of the investigator is to select the specific one of the collection of techniques to be used in any particular problem.

PURPOSES FOR CLUSTERING

Before considering more usual examples of cluster analysis, it is useful to consider a simple example presented by Anderberg[3] to show that different groupings are appropriate for different purposes. Anderberg uses the 52 entities in a deck of playing cards for his example. One possible grouping of these cards into two groups is shown in Figure 8–1a, where the clusters are the red cards and the black ones. Anderberg suggests that such a clustering might reflect a small child's perception of the cards. Figure 8–1b shows an alternative grouping into two clusters, one containing all face cards and the other containing the remaining cards. This clustering is particularly meaningful to a bridge player in determining the point count of a hand.

Figure 8–1c shows another grouping, with 13 groups, one for each face value. Figure 8–1d shows 26 groups, with one for each face value and color combination. Figure 8–1e shows 4 groups, one for each suit. Finally, Figure 8–1f shows the clustering of most significance for players of the game hearts.

No one of these groupings can be called the correct one. Instead,

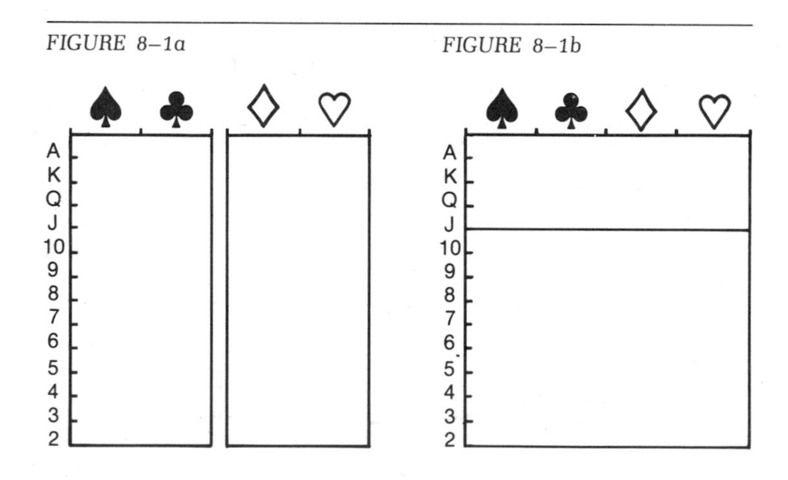

FIGURE 8–1a

FIGURE 8–1b

[2]The techniques go by a variety of names, in a variety of fields, including cluster analysis, numerical taxonomy, and pattern recognition.

[3]Michael R. Anderberg, *Cluster Analysis for Applications*, (New York: Academic Press, 1973).

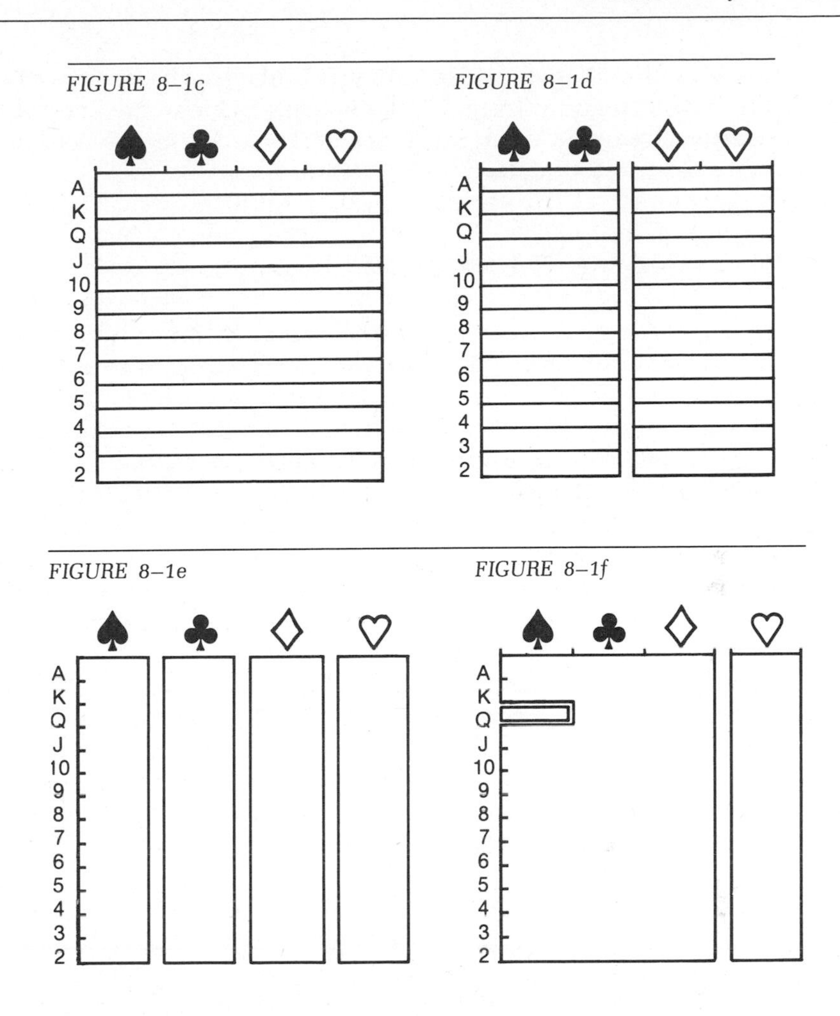

FIGURE 8–1c

FIGURE 8–1d

FIGURE 8–1e

FIGURE 8–1f

each of these choices (and others not shown here) is meaningful to certain people under certain circumstances. In general, cluster analysis is situation-specific and, often, exploratory. The results of a particular cluster analysis must be interpreted in the context of a particular situation and a particular purpose for analysis. Therefore, the choice of technique depends on the context and purpose.

SEPARATION BETWEEN VERSUS SIMILARITY WITHIN

Often, the objective of clustering is to produce one of two conditions:

1. Groups which are clearly separated from one another.
2. Groups whose members are very similar.

These two purposes are not the same and may well conflict with each other. In addition, the terms "clearly separated" and "very similar" are not precisely defined and it is not always obvious what they mean in specific problems. A series of examples in which we consider clustering observations will illustrate these points. In these examples, each object (or observation) is assumed to be characterized by values of two variables, x_1 and x_2. The observations are then represented in simple two-dimensional graphs with the horizontal dimension corresponding to x_1 and the vertical corresponding to x_2. A point is shown at the x_1 and x_2 values for each observation. Dissimilarity between a pair of objects is taken to be the ordinary distance between the points representing the two observations. In a situation in which we were studying consumption patterns, for example, x_1 might be annual spending on food and x_2 spending on clothing.

The first example is given in Figure 8–2a. Here the two clusters are very homogenous; no two points within one cluster are very far apart. The clusters are also clearly separated from one another. We would expect just about anyone who looked at the figure to identify the two clear clusters.

Figure 8–2b is a bit less clear. Many observers would identify two clear clusters, clearly separated from one another. The two clusters satisfy the separation-between criterion well. They would not necessarily satisfy the similarity-within criterion, however. Notice that some points in the left cluster (those at the right boundary of that cluster) are closer to some points in the other cluster than they are to the points at the opposite side of their own cluster. Figure 8–2c presents a more extreme example of the same phenomenon. There are two well-separated but nonhomogeneous clusters.

Figure 8–2d shows an example in which there would appear to be only one group, under either criterion. In Figure 8–2e, in contrast, there are not clearly separated clusters, yet many observers would identify two clusters under the similarity-within criterion; each such cluster is roughly round. The two apparent clusters lie very close to one another.

Figure 8–2f shows the points from Figure 8–2a with a few stray points added. The extra points might be considered noise. Now, in any strict sense, the figure shows neither separation between nor similarity within—unless we exclude some points (as noise) and concentrate only on the rest. Figure 8–2g shows a similar situation, where without noise we can see two clearly separated (though not homogeneous) clusters; with noise we cannot.

Dealing in only two dimensions, investigators can judgmentally identify situations such as these and can make sensible intuitive groupings of points into clusters satisfying one of the two criteria, if such clusters exist. They can label stray points as noise and ignore the noise

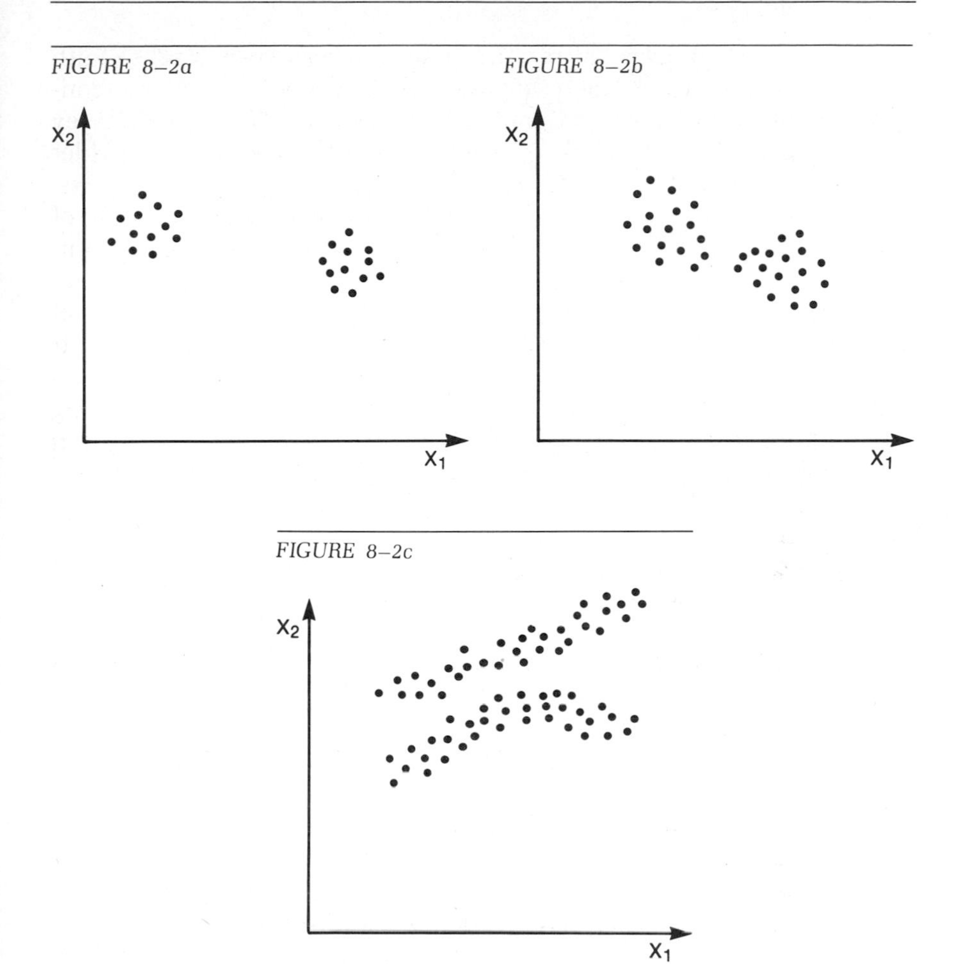

FIGURE 8–2a

FIGURE 8–2b

FIGURE 8–2c

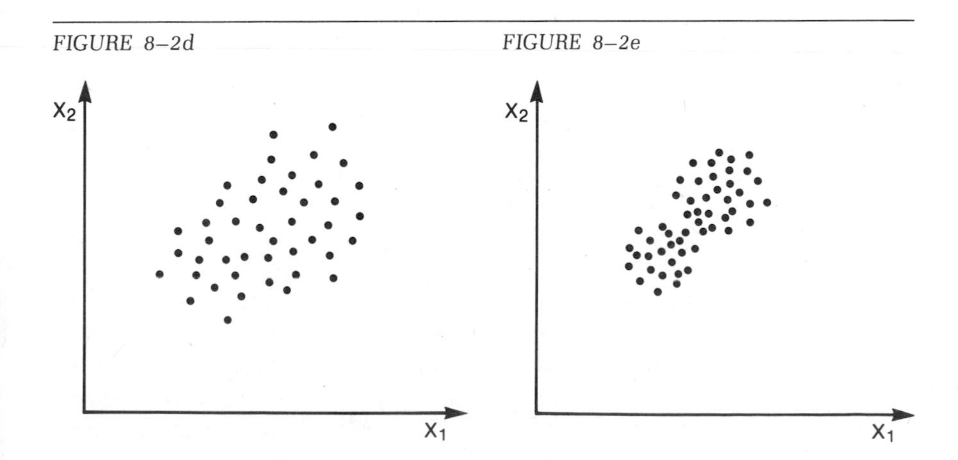

FIGURE 8–2d

FIGURE 8–2e

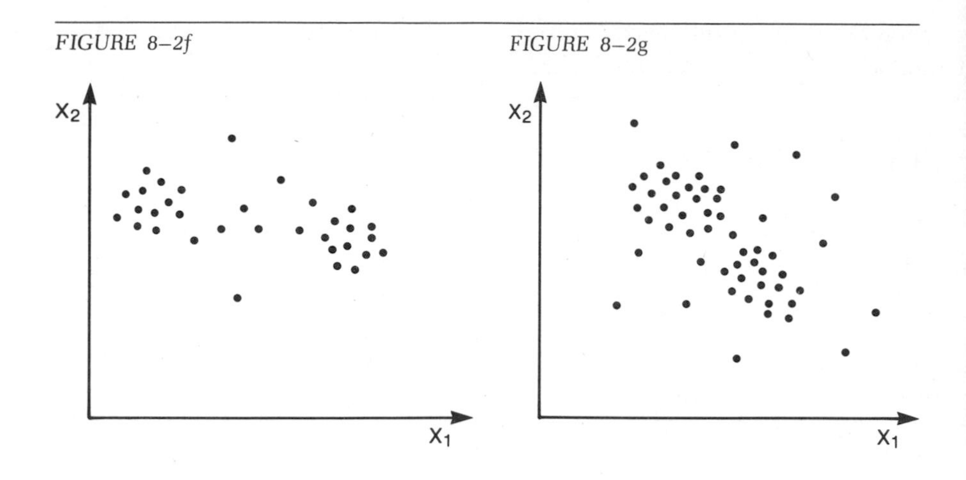

in their analyses where appropriate. Even in such relatively straightforward cases, the investigator must still determine, from the characteristics of the particular problem at hand, what is the general purpose of the grouping, but after that first important step the investigation of two-dimensional plots is fairly easy and can be performed without a computer program.

Problems arise in more than two dimensions, the usual case in clustering. If investigators want to consider more than two dimensions (and possibly a rather large number of dimensions) for describing each of a set of objects and then to group those objects into clusters, visualization becomes considerably less practical. Suppose that we want to use five variables or descriptors. With one dimension per variable, we would want to measure dissimilarity by distances in five-dimensional space. We cannot draw or visualize the situation and, therefore, we will in general try to find more mechanical ways of grouping the objects. The mechanical methods, or algorithms, can then be carried out by computer.

Looking back over the examples in Figure 8–2, the reader will see that a variety of intuitive, or heuristic, techniques were used to identify clusters. All of the numerous algorithms for performing cluster analysis can be considered alternative ways of trying to make mechanical and explicit the types of procedures used in a less formal way in perceiving the examples in Figure 8–2.

In using cluster analysis the user must specify both a measure of dissimilarity (or similarity) and a method of grouping. This chapter considers the choice of method first. In doing so we will use the simple distance measure used in the examples of Figure 8–2. In other words, we will use ordinary, or Euclidean, distance (or, more accurately, distance squared) to reflect dissimilarity.[4] Various distance measures are considered briefly later in the chapter.

FULL ENUMERATION

The reader may wonder why computer programs for cluster analysis do not follow the easy-to-imagine procedure of considering all possible clusterings of a group of points, considering all possible groupings into each possible number of groups. The program could then select the configurations that were best under some specified measure of goodness.

Unfortunately, such full enumeration is in general not feasible. Even for rather small numbers of points, numbers of possible clusterings rapidly become astronomical and soon exceed computer capabilities. Therefore, we are forced to use algorithms which consider only a few of the total number of possible groupings. Algorithms for cluster analysis are intended to use sensible ways to select groupings for consideration.

THE NEAREST-NEIGHBOR ALGORITHM

The first practical method for grouping which we will consider is an example of what is called an *agglomerative* method. The name arises from the fact that such algorithms begin with the objects all considered to be individual points and proceed to build up clusters by joining points successively into groups.

In order to be able to follow all of the calculations involved, we will use an extremely small example, with only five points and two variables, or dimensions. Each point might be an individual and the two variables might give the individual's weekly consumption of beer (cans) and wine (glasses), respectively. Table 8–1 shows the values of the two variables for each point. Figure 8–3 is a plot of the corresponding points. Table 8–2 lists the squared Euclidean distance between each pair of points.[5]

---

[4]In two dimensions, Euclidean distances have the normal everyday meaning of distance; squared distances are simply those distance values squared. In more than two dimensions the operational definition of the squared Euclidean distance between two points is: Find the difference between the points on the first dimension and square it. Similarly, square the difference on the second dimension. Proceed to find squared differences for each dimension. The sum of these squares is the squared Euclidean distance.

[5]For example, the squared distance between b and c is $(1 - 8)^2 + (2 - 2)^2 = 49 + 0 = 49$.

TABLE 8–1

Point	X_1 (Beer)	X_2 (Wine)
a	1	1
b	1	2
c	8	2
d	6	3
e	8	0

FIGURE 8–3

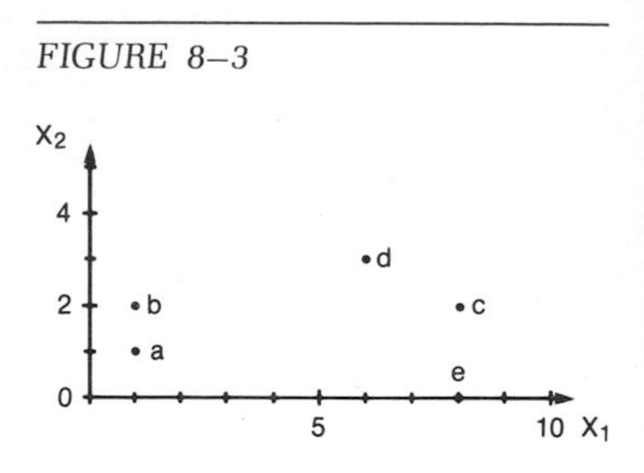

The *nearest-neighbor algorithm* begins by joining the closest points. In this example, *a* and *b* (with squared distance 1) are joined. Figure 8–4 shows the result.

The nearest-neighbor method next calls for examining the distances between all entities (groups and ungrouped points) and merging the closest two.[6] To follow this procedure, we must somehow define the squared distance between a group (*ab*) and a point (such as *c* or *d*). In the nearest-neighbor procedure, the distance between two entities is defined as the shortest of the distances between one point in one entity and one point in the other. Hence the name nearest neighbor. For example, the distance between the point *d* and the group *ab* is 26—the smaller of two distances, that between *d* and *a* (29) and that between *d* and *b* (26). Table 8–3 shows the new distance matrix. Points *c* and *e* are now the closest entities and they are joined, as shown in Figure 8–5.

The next distance matrix is shown in Table 8–4. The closest entities are now point *d* and group *ce* which are merged at the next step, giving Figure 8–6.

TABLE 8–2

Squared Distances

	a	*b*	*c*	*d*	*e*
a	0	1	50	29	50
b	1	0	49	26	53
c	50	49	0	5	4
d	29	26	5	0	13
e	50	53	4	13	0

FIGURE 8–4

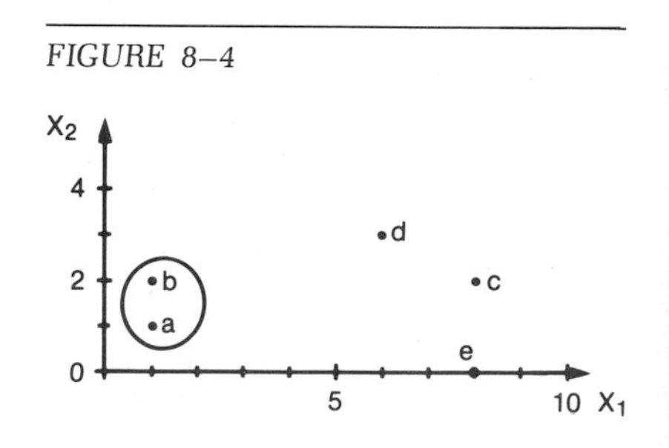

[6] If there is a tie, both mergers can be made. (Alternatively, some versions of the procedure might choose one of the mergers in case of a tie.)

TABLE 8–3

Squared Distances

	ab	c	d	e
ab	0	49	26	50
c	49	0	5	4
d	26	5	0	13
e	50	4	13	0

FIGURE 8–5

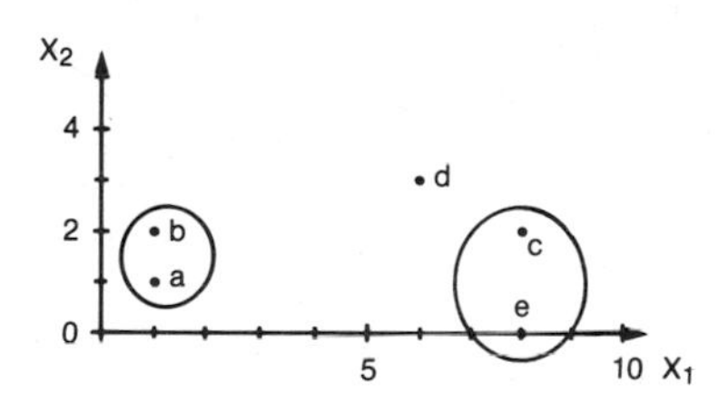

TABLE 8–4

Squared Distances

	ab	d	ce
ab	0	26	49
d	26	0	5
ce	49	5	0

FIGURE 8–6

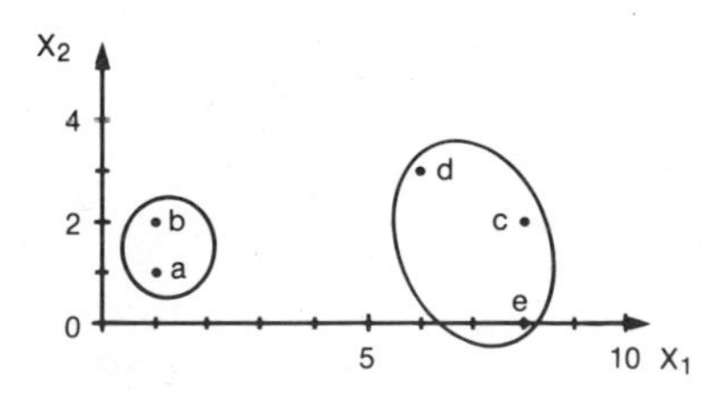

The resulting distances are given in Table 8–5. The last step is to join *ab* with *dce*, as shown in Figure 8–7.

This procedure, like most agglomerative techniques, started with the individual points and proceeded until all the points had been merged into one cluster. The basic idea, however, is that one (or more) of the intermediate stages of the procedure are likely to have the most meaningful groupings. In the very simple example under study, we can just look at the figures and choose a grouping—such as that shown in Figure 8–6. Recall, however, that the whole purpose of considering formal clustering methods was to allow analysis of problems in more than two dimensions. (The current extremely simple example is being used only for illustrative purposes.) Hence, we should imagine that we do not have available Figures 8–3 through 8–7 and that we must somehow decide on a cluster configuration from numerical results, such as those in Tables 8–1 through 8–5.

The most common way for such results to be described is with a branching diagram called a *dendrogram*. Figure 8–8a gives a dendrogram for the nearest-neighbor clustering that was just performed. The scale along the left side of the diagram gives squared distance values. The horizontals are drawn at the squared distances that were smallest at the time of the merger between the entities shown (in other words, 1 for *a* and *b*, 4 for *c* and *e*, 5 for *d* and *ce*, and 26 for *ab* and *dce*). As we move upward from the bottom of the dendrogram, we find the mergers in the same order they occurred. (Note that, to achieve this effect, it was

TABLE 8–5

Squared Distances

	ab	*dce*
ab	0	26
dce	26	0

FIGURE 8–7

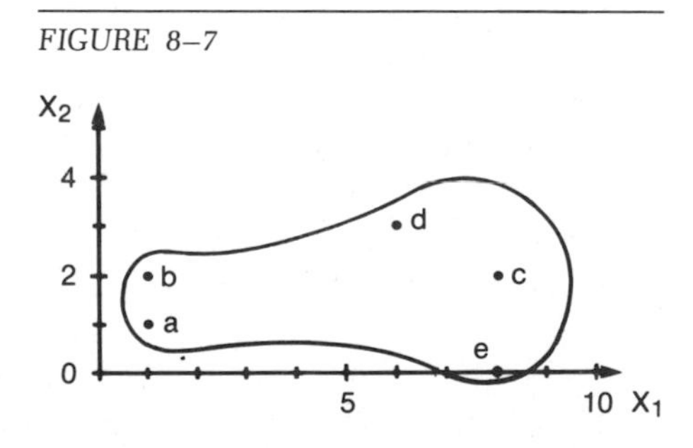

FIGURE 8–8a

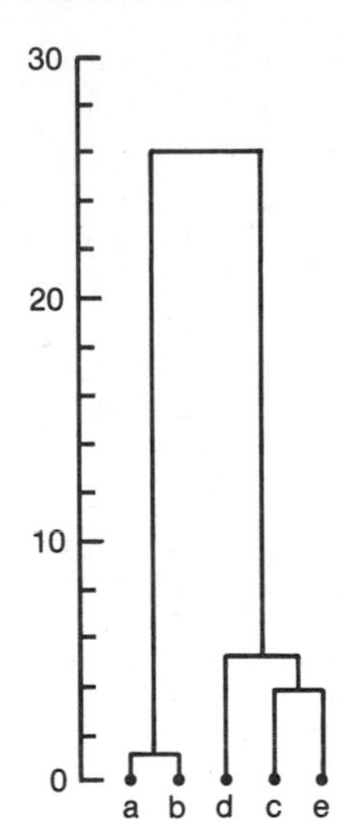

FIGURE 8–8b

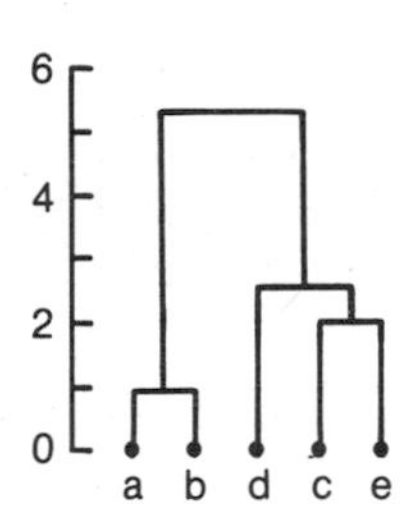

necessary to change the order of the points from the natural alphabetic ordering, putting *c* and *e* next to one another.)

Figure 8–8b gives similar information but uses a scale of Euclidean distance rather than squared Euclidean distance.[7] In general, as is shown by Figure 8–8, Euclidean distances and squared Euclidean distances give the same order of mergers; only the scales in the resulting dendrograms differ.

In interpreting dendrograms we usually interpret a large jump in scale as showing a merger of two entities which are not naturally grouped together. In Figure 8–8 (both parts) the large jumps occur for the merger of *ab* with *dce*. Hence, we would generally conclude that such a merger should not take place and that the data show two clusters. (In doing so, we would arrive at the same conclusions suggested by Figures 8–3 through 8–7.)

Table 8–6 lists another set of points, each characterized by its values on two variables. Figure 8–9 shows the dendrogram (in terms of squared distances) that results from nearest-neighbor clustering. Notice that more than one merger can occur at the same level. The scale values jump at the last merger. Therefore, we would guess that points *bdehacgi* form one cluster and *fjlkmnop* form the other. Plotting the points (in Figure 8–10) confirms this guess.

[7]Euclidean distances are simply the square roots of squared Euclidean distances.

TABLE 8–6

Point	X_1	X_2
a	23	15
b	19	14
c	24	13
d	18	12
e	21	12
f	6	10
g	24	10
h	17	9
i	22	9
j	8	8
k	11	7
l	6	6
m	9	5
n	9	3
o	12	3
p	6	2

FIGURE 8–9

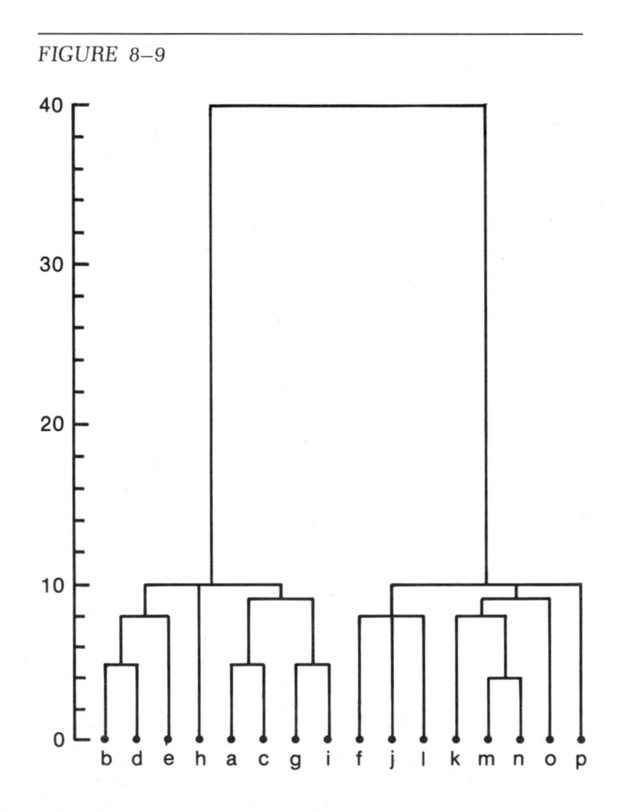

FIGURE 8–10

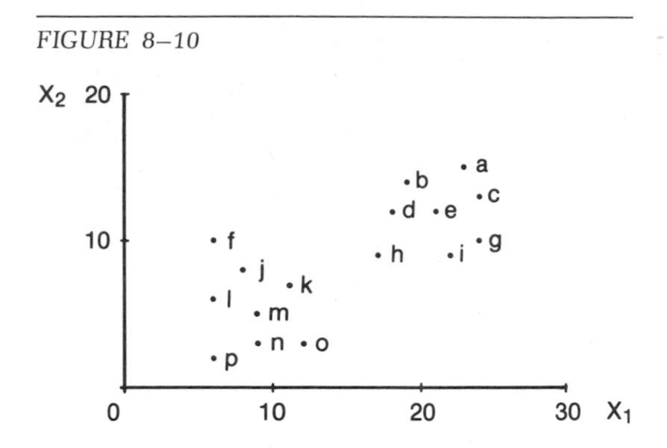

FARTHEST-NEIGHBOR ALGORITHM

Like nearest neighbor, the *farthest-neighbor algorithm* is an agglomerative procedure for clustering. As did nearest neighbor, farthest neighbor begins by merging the two closest points to form a single entity. In the simple example from Table 8–1, points *a* and *b* are merged. At this point, however, the two procedures diverge in the ways in which they define the distance between two entities. The farthest-neighbor procedure defines such a distance as the largest pairwise distance between a point in one entity and a point in the other. For the current example, the distance between the point *d* and the cluster *ab* is 29—the larger of two distances, that between *d* and *a* (29) and that between *d* and *b* (26). The new (squared) distances are shown in Table 8–7.

Next, the closest two entities (*c* and *e*) are joined. The new distances are shown in Table 8–8. Then, the closest entities are *ce* and *d*, which are merged to give the values in Table 8–9. Finally, *ab* and *dce* are merged. The resulting dendrogram is shown in Figure 8–11. The order of mergers for this simple example turns out to be the same for nearest neighbor and for farthest neighbor, although such is not generally the case.

TABLE 8–7

Squared Distances

	ab	*c*	*d*	*e*
ab	0	50	29	53
c	50	0	5	4
d	29	5	0	13
e	53	4	13	0

TABLE 8–8

	ab	d	ce
ab	0	29	53
d	29	0	13
ce	53	13	0

TABLE 8–9

	ab	dce
ab	0	53
dce	53	0

FIGURE 8–11

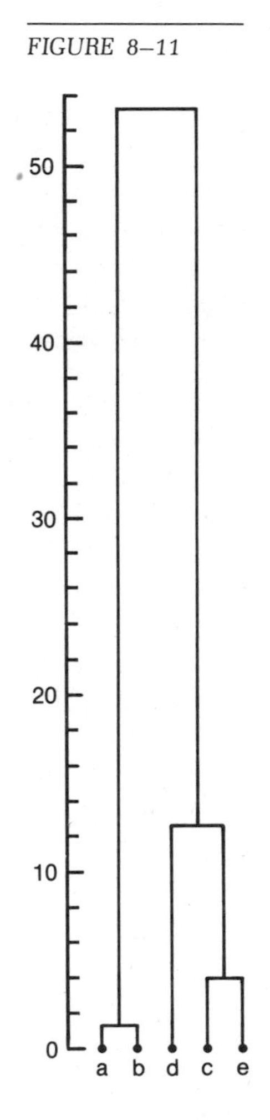

The dendrograms differ, however, showing the different squared distances at which mergers occurred. The figure for farthest neighbor gives the largest distance between pairs of points within the merged entity. (Notice that the merger of *d* with *ce* is shown at a higher scale value for farthest neighbor than for nearest neighbor.) Thus, farthest neighbor tends to highlight relative lack of similarity within clusters.

The *dce* cluster is not as tight as is the *ab* one, and the farthest-neighbor dendrogram highlights this point.

In general, the farthest-neighbor algorithm is particularly suited to finding homogeneous clusters under the similarity-within objective. Since it considers the largest pairwise distance between points in two entities, it is reluctant to put into the same group any two points that are not very similar to one another. Farthest neighbor is therefore good at identifying rather compact groups with approximately the same diameter (or distance across them), as for example in Figure 8–12.

The farthest neighbor method is less affected by the presence of noise than are other techniques. In Figure 8–2f, for example, the noise points between the two clusters would lead nearest neighbor to link the two together readily; several noise points are near other noise points and/or near members of the clusters and nearest neighbor would make them into a bridge. By constrast, farthest neighbor would be reluctant to join the clusters despite the noise; the largest pairwise distance between points in the two clusters is large.

On the other hand, farthest neighbor is poorly suited for finding naturally separated but nonhomogeneous clusters. Figure 8–13 gives an example. Nearest neighbor would easily find the two clusters, but farthest neighbor would not.

Thus, compared with farthest neighbor, nearest neighbor is well suited for finding clearly separated natural groupings (the separation-between objective). It is affected particularly badly by the presence of noise. Especially in the presence of noise, nearest neighbor is likely to display *chaining*, putting together a long linkage of entities in a highly nonhomogeneous group. (On the other hand, the same tendency is what makes nearest neighbor find the groups in Figure 8–13.) Nearest neighbor is very poor at finding homogeneous groups.

FIGURE 8–12

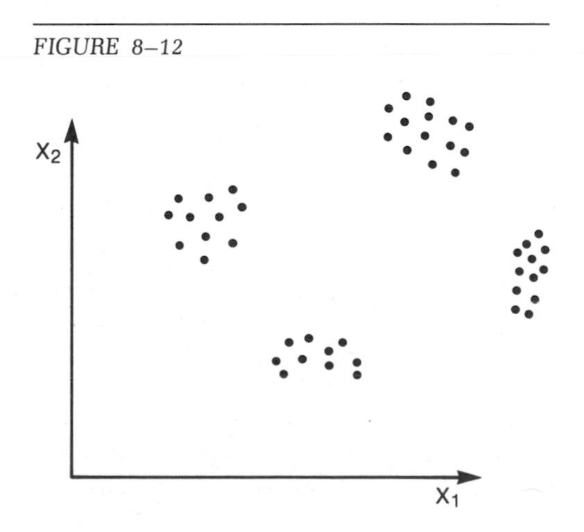

FIGURE 8–13

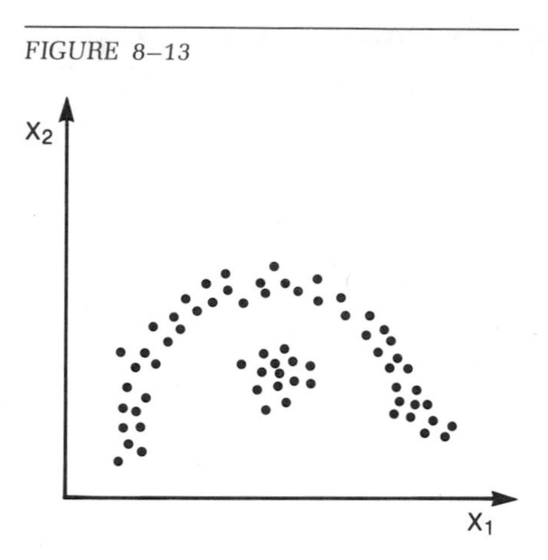

MINIMUM-SQUARED-ERROR METHOD

The *minimum-squared-error method* is another technique for agglomerative clustering. The technique involves the notion of the *centroid* of a group of points. The centroid is also a point; it is the generalization to two or more dimensions of the mean or average. In two dimensions, the centroid's value on the first dimension is the average first-dimension value of all the points in the group; its value on the second dimension is the average second-dimension value. The definition generalizes to higher dimensions.

The basic merging procedure in the minimum-squared-error method is to select at each stage that merger of two entities that gives the smallest total squared distance between the individual points and the centroids of the entities to which they are assigned after the merger. (Minimizing total squared distance in this definition is equivalent to minimizing average squared distance.) This procedure is best illustrated with an example.

For the set of points in Table 8–1, Table 8–10 gives an evaluation of all possible first mergers. Notice that each point in the set belongs to a group with a centroid. If there is only one point in a group, the centroid is simply the same as the point.

The table shows that a minimum squared distance after merger occurs for the joining of points *a* and *b*; therefore, these two points are merged. Then, in the second step of this procedure, there are six possible mergers, as shown in Table 8–11.

Joining *c* and *e* gives the smallest total squared distance after the merger and therefore those two points are joined. Next, Table 8–12 shows the possible mergers for the third step of the process.

TABLE 8–10

Possible Step	Point	Centroid	Squared Distance
A. Merge a and b (1,1) and (1,2):	a	(1,1.5)	$.5^2$
	b	(1,1.5)	$.5^2$
	c	(8,2)	0
	d	(6,3)	0
	e	(8,0)	0
			.5 = Total squared distance
B. Merge a and c (1,1) and (8,2):	a	(4.5,1.5)	$3.5^2 + .5^2 = 12.5$
	b	(1,2)	0
	c	(4.5,1.5)	12.5
	d	(6,3)	0
	e	(8,0)	0
			25
C. Merge a and d (1,1) and (6,3):	a	(3.5,2)	$2.5^2 + 1^2 = 7.25$
	b	(1,2)	0
	c	(8,2)	0
	d	(3.5,2)	7.25
	e	(8,0)	0
			14.5
D. Merge a and e (1,1) and (8,0):	a	(4.5,.5)	$3.5^2 + .5^2 = 12.5$
	b	(1,2)	0
	c	(8,2)	0
	d	(6,3)	0
	e	(4.5,.5)	12.5
			25
E. Merge b and c (1,2) and (8,2):	a	(1,1)	0
	b	(4.5,2)	$3.5^2 = 12.25$
	c	(4.5,2)	12.25
	d	(6,3)	0
	e	(8,0)	0
			24.5
F. Merge b and d (1,2) and (6,3):	a	(1,1)	0
	b	(3.5,2.5)	$2.5^2 + .5^2 = 6.5$
	c	(8,2)	0
	d	(3.5,2.5)	6.5
	e	(8,0)	0
			13

TABLE 8–10 (concluded)

Possible Step	Point	Centroid	Squared Distance
G. Merge *b* and *e* (1,2) and (8,0):	*a*	(1,1)	0
	b	(4.5,1)	$3.5^2 + 1^2 = 13.25$
	c	(8,2)	0
	d	(6,3)	0
	e	(4.5,1)	13.25
			26.5
H. Merge *c* and *d* (8,2) and (6,3):	*a*	(1,1)	0
	b	(1,2)	0
	c	(7,2.5)	$1^2 + .5^2 = 1.25$
	d	(7,2.5)	1.25
	e	(8,0)	0
			2.5
I. Merge *c* and *e* (8,2) and (8,0):	*a*	(1,1)	0
	b	(1,2)	0
	c	(8,1)	$1^2 = 1$
	d	(6,3)	0
	e	(8,1)	1
			2
J. Merge *d* and *e* (6,3) and (8,0):	*a*	(1,1)	0
	b	(1,2)	0
	c	(8,2)	0
	d	(7,1.5)	$1^2 + 1.5^2 = 3.25$
	e	(7,1.5)	3.25
			6.5

Merging *d* with *ce* gives the smallest total squared distance and that step is taken. Finally, *ab* and *dce* are joined. The results of this procedure could be displayed in a dendrogram, as were the results of the earlier procedures, although the dendrogram is not included here.

Because the minimum-squared-error procedure includes a squared-error criterion, it should not be surprising that the program is extremely reluctant to include outliers in clusters. Therefore, the procedure is relatively good in terms of the similarity-within objective and is less affected than is nearest neighbor by the presence of noise.[8] On the other hand, this technique is generally not appropriate when the investigator wants to find clearly separated natural clusters such as those shown in Figure 8–13.

[8]The minimum-squared-error method is especially suited for finding clusters that are approximately round or spherical and approximately equal in size.

TABLE 8–11

Possible Step	Point	Centroid	Squared Distance
A. Merge *c* with *ab*			
(8,2), (1,1), (1,2):	*a*	(10/3,5/3)	$(7/3)^2 + (2/3)^2 = 53/9$
	b	(10/3,5/3)	$(7/3)^2 + (1/3)^2 = 50/9$
	c	(10/3,5/3)	$(14/3)^2 + (1/3)^2 = 197/9$
	d	(6,3)	0
	e	(8,0)	0
			$300/9 = 33.33$
B. Merge *d* with *ab*			
(6,3), (1,1), (1,2):	*a*	(8/3,2)	$(5/3)^2 + 1^2 = 34/9$
	b	(8/3,2)	$(5/3)^2 + 0 = 25/9$
	c	(8,2)	0
	d	(8/3,2)	$(10/3)^2 + 1^2 = 109/9$
	e	(8,0)	0
			$168/9 = 18.67$
C. Merge e with *ab*			
(8,0), (1,1), (1,2):	*a*	(10/3,1)	$(7/3)^2 = 49/9$
	b	(10/3,1)	$(7/3)^2 + 1^2 = 58/9$
	c	(8,2)	0
	d	(6,3)	0
	e	(10/3,1)	$(14/3)^2 + 1^2 = 205/9$
			$312/9 = 34.67$
D. Merge c and d			
{(8,2), (6,3)} {(1,1), (1,2)}:	*a*	(1,1.5)	$.5^2 = .25$
	b	(1,1.5)	$.5^2 = .25$
	c	(7,2.5)	$1^2 + .5^2 = 1.25$
	d	(7,2.5)	$1^2 + .5^2 = 1.25$
	e	(8,0)	0
			3.0
E. Merge c and e			
{(8,2), (8,0)} {(1,1), (1,2)}:	*a*	(1,1.5)	$.5^2 = .25$
	b	(1,1.5)	$.5^2 = .25$
	c	(8,1)	$1^2 = 1$
	d	(6,3)	0
	e	(8,1)	$1^2 = 1$
			2.5
F. Merge d and e			
{(6,3), (8,0)} {(1,1), (1,2)}:	*a*	(1,1.5)	$.5^2 = .25$
	b	(1,1.5)	$.5^2 = .25$
	c	(8,2)	0
	d	(7,1.5)	$1^2 + 1.5^2 = 3.25$
	e	(7,1.5)	$1^2 + 1.5^2 = 3.25$
			7.0

TABLE 8–12

Possible Step	Point	Centroid	Squared Distance
A. Merge *d* with *ab*:	*a*	(8/3,2)	$(5/3)^2 + 1^2 = 34/9$
	b	(8/3,2)	$(5/3)^2 = 25/9$
	c	(8,1)	$(10/3)^2 + 1^2 = 109/9$
	d	(8/3,2)	1^2
	e	(8,1)	1^2
			186/9 = 20.67
B. Merge *d* with *ce*:	*a*	(1,1.5)	$.5^2 = .25$
	b	(1,1.5)	$.5^2 = .25$
	c	(22/3,5/3)	$(2/3)^2 + (1/3)^2 = 5/9$
	d	(22/3,5/3)	$(4/3)^2 + (4/3)^2 = 32/9$
	e	(22/3,5/3)	$(2/3)^2 + (5/3)^2 = 29/9$
			7.83
C. Merge *ab* with *ce*:	*a*	(4.5,1.25)	$3.5^2 + .25^2 = 12.31$
	b	(4.5,1.25)	$3.5^2 + .75^2 = 12.81$
	c	(4.5,1.25)	$3.5^2 + .75^2 = 12.81$
	d	(6,3)	0
	e	(4.5,1.25)	$3.5^2 + 1.25^2 = 13.81$
			51.74

HILL-AND-VALLEY METHODS

Hill-and-valley methods are a second classification of clustering techniques, different from the agglomerative methods. Hill-and-valley techniques are intended for use in situations in which investigators believe there are well-defined natural clusters in data but that the clusters are not clearly separated, often because of the presence of noise (or, possibly, simply because the clusters are close to one another). In other words, such methods are appropriate for analysis with the similarity-within objective, especially in the presence of noise.

Figures 8–2f and 8–2g showed examples of clusters somewhat obscured by what may be noise points. To extract the underlying clusters for situations such as those, the hill-and-valley procedures involve a concept of *density* of points. The concept rests on the premise that noise points will tend to be less densely packed than will member points of true clusters and, therefore, that a useful approach for finding the true clusters is to look for points in regions which have denser concentrations of points. For both Figure 8–2f and Figure 8–2g, the true clusters are in the densest collections of points and the techniques based on density should be effective.

To implement density-based techniques on computers, we need to have an operational definition of density that does not involve visual inspection of the points. A common measure of density involves the simple idea that if a point is densely surrounded by other points, there will be many points close to it. Therefore, distances to neighboring points are a useful basis for measuring density. To be more precise, we really want a measure that requires several nearby points. For that reason, the common measure for the density of a point is defined in terms of the distance from that point to its kth nearest neighbor, with k an integer specified by the user. For example, k might be 5, so that we would measure for each point under consideration the distance between that point and the fifth-nearest other point. Points with smaller fifth-nearest-neighbor distances would then be considered more dense than would points with larger such distances.

The two algorithms for hill-and-valley clustering considered here are alike in that they use the kth-nearest-neighbor definition of density and, in addition, that they consider points in order of decreasing density, either using the points to start new clusters or else merging them into existing clusters. The details of the procedures are different, however.

The first such procedure, called *unimodal*, first identifies key points to become the nuclei used in forming clusters. The nuclei are defined as points which are more dense than any of their neighbors. (In other words, they are local maxima of the density function.) After all of the cluster nuclei have been identified, the algorithm proceeds to consider each of the remaining points in turn, in order of decreasing density. Each point is merged into the existing cluster nearest to it (in the nearest neighbor sense). Such additional points are never used to start new clusters. Thus, the number of clusters is determined at the start as the number of nuclei found. After that step, the number of clusters remains unchanged, with other points added to one or another of the existing clusters.

This unimodal procedure as just described does not end with a single cluster, as did the agglomerative methods described earlier.[9] It may be useful to proceed with clustering to force mergers until only one cluster remains. If so, some agglomerative technique, such as the nearest neighbor algorithm, can be used to accomplish the additional clustering.

Figure 8–14 (a and b) shows plots of the steps of the unimodal algorithm. The first part of the figure shows an initial configuration of points, which seems to have three clear clusters plus noise. The second part of the figure shows three cluster nuclei, each identified by an asterisk rather than a dot. Such points might have been identified according to their fifth-nearest-neighbor distances, for example. It also shows the results after the unimodal clustering (but before the additional nearest-neighbor linkages to reduce the figure to one cluster). The noise points

[9]Recall that each of the agglomerative methods started with the maximum number of groups (each consisting of a single point) and then performed mergers until all entities coalesced into a single one.

FIGURE 8–14a

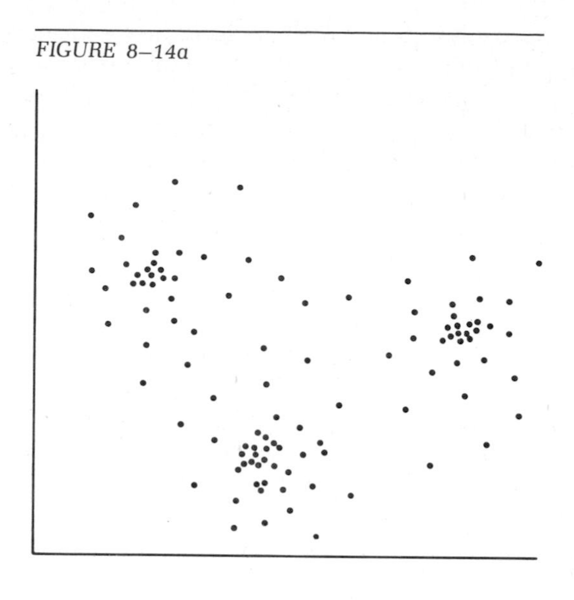

FIGURE 8–14b

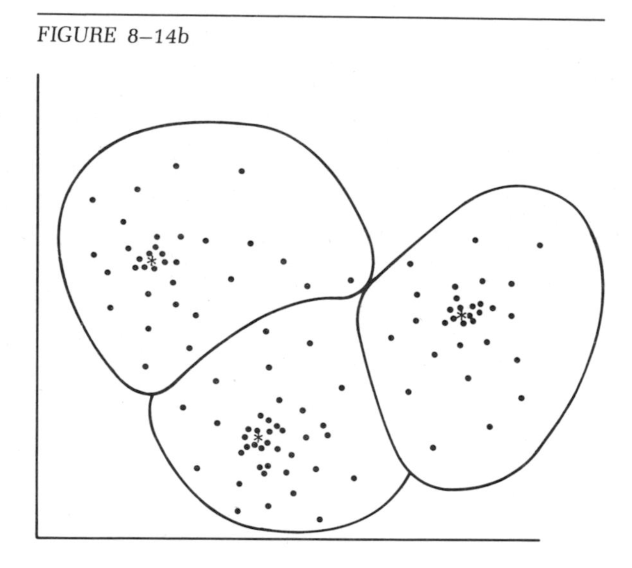

have each been added to a cluster, but the presence of those points has not obscured the existence of the three underlying groupings.

An alternative hill-and-valley method is called the *progressive-threshold* method. That technique begins by finding the distance to the kth-nearest-neighbor for each point, using a value of k specified by the investigator. It then considers the points one by one in order, starting with the point with the smallest k^{th}-nearest-neighbor distance and proceeding to the point with largest such distance. At each step, the procedure works only with the current point and the points preceding it in the ordering. On each such step, the distance value for the current point serves as a threshold distance (hence the name progressive threshold). The procedure merges any two entities (from among the entities considered so far) only if those two entities are nearer to one another, in a nearest-neighbor sense, than the current threshold value. In other words, the procedure considers only entities formed previously in the algorithm and does not consider points that are less dense than the one under current consideration. For those previously formed entities, the procedure performs some steps of the nearest-neighbor algorithm but does not proceed to merge any entities which are farther from one another than the current threshold. Any two clusters that are closer than the current threshold will be merged. After all indicated mergers have been performed, the algorithm proceeds to the next point, establishing a new threshold and performing whatever nearest-neighbor mergers are appropriate. It then continues through the list of points, in order of density. As described, the procedure will end with a few (or perhaps only one) cluster. If more than one cluster remains at the end, additional clustering with the pure nearest-neighbor procedure would likely be desirable to merge the remaining clusters into one.

In general practice, the user must specify a value for k, for either the unimodal or the progressive-threshold procedure. The value will have to depend both on the number of clusters anticipated and on the number of points involved. In particular, k should be small compared with the number of points expected per cluster. Values of 5 to 10 are typical, although ks larger than 10 can be used with large numbers (hundreds) of points and ks less than 5 are sometimes used when the user is worried about having too few points.

Figure 8–15 (a and b) shows a plot of a simple example to demonstrate the progressive-threshold method. The first part of the figure shows only the points; the second shows points and labels. The figure contains few points and should be considered for illustrative purposes only. A value of 3 has been used for k, again for illustrative purposes. Table 8–13 lists the points in the data and for each gives the squared

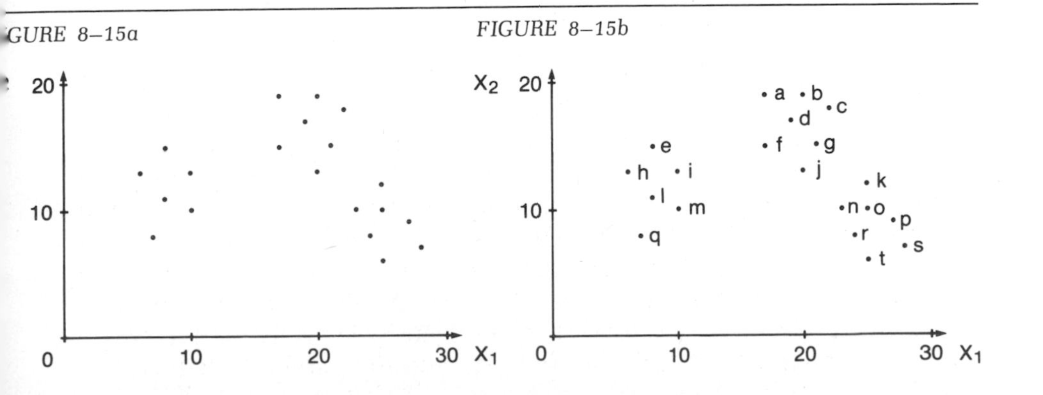

TABLE 8–13

Point	X_1	X_2	Name of Third Nearest Neighbor	Squared Distance to Third Nearest Neighbor
a	17	19	*f*	16
b	20	19	*a*	9
c	22	18	d or g (tie)	10
d	19	17	*a*, *f*, or g (tie)	8
e	8	15	*l*	16
f	17	15	*a* or g (tie)	16
g	21	15	c	10
h	6	13	*i*	16
i	10	13	m	9
j	20	13	d	17
k	25	12	p	13
l	8	11	*h* or *i* (tie)	8
m	10	10	q	13
n	23	10	k	8
o	25	10	r or p (tie)	5
p	27	9	r	10
q	7	8	h	26
r	24	8	n, o, or t (tie)	5
s	28	7	r	17
t	25	6	p	13

distance to the third-nearest neighbor. Finally, Figure 8–16 gives the dendrogram for the progressive threshold clustering of these data. The scale values in the dendrogram give the progressive threshold in effect at each stage. The first threshold is 5, the smallest number in the last column in Table 8–13. Points *o* and *r* both have 5 as their squared third-neighbor distance, so *o* and *r* enter. Since *o* and *r* are within 5 units (squared) of one another, they merge. The next threshold is 8, the next smallest entry in the last column in Table 8–13. Points *d*, *l*, and *n* have squared third-neighbor distances of 8, so they enter. Points *d* and *l* do not merge with anything at this stage. Point *n* enters and merges with entity *or*. At the next threshold, 9, points *b* and *i* enter. Point *b* merges with *d* and *i* joins *l*. At the threshold of 10, *c*, *g*, and *p* enter. *p* joins *nor*, while *c* and *g* both join *bd*. The next threshold is 13, with *k*, *m*, and *t* entering. *k* and *t* join *pnor*, while *m* joins *il*. At the threshold of 16, *a*, *e*, *f*, and *h* enter. *h* and *e* join *ilm*, while *a* and *f* join *cbdg*. At 17, *j* and *s* enter; *j* joins *facbdg*, while *s* joins *kpnort*. The next threshold is 26. *q* enters and joins *heilm*. Also, *facbdgj* and *skpnort* are within this distance of one another. (The squared distance between *j* and *n* is 18.) Therefore, the two clusters merge. There are no more thresholds left. The last merger is accomplished as a simple nearest-neighbor merger at a squared distance of 53 (the squared distance between *i* and *f*).

FIGURE 8–16

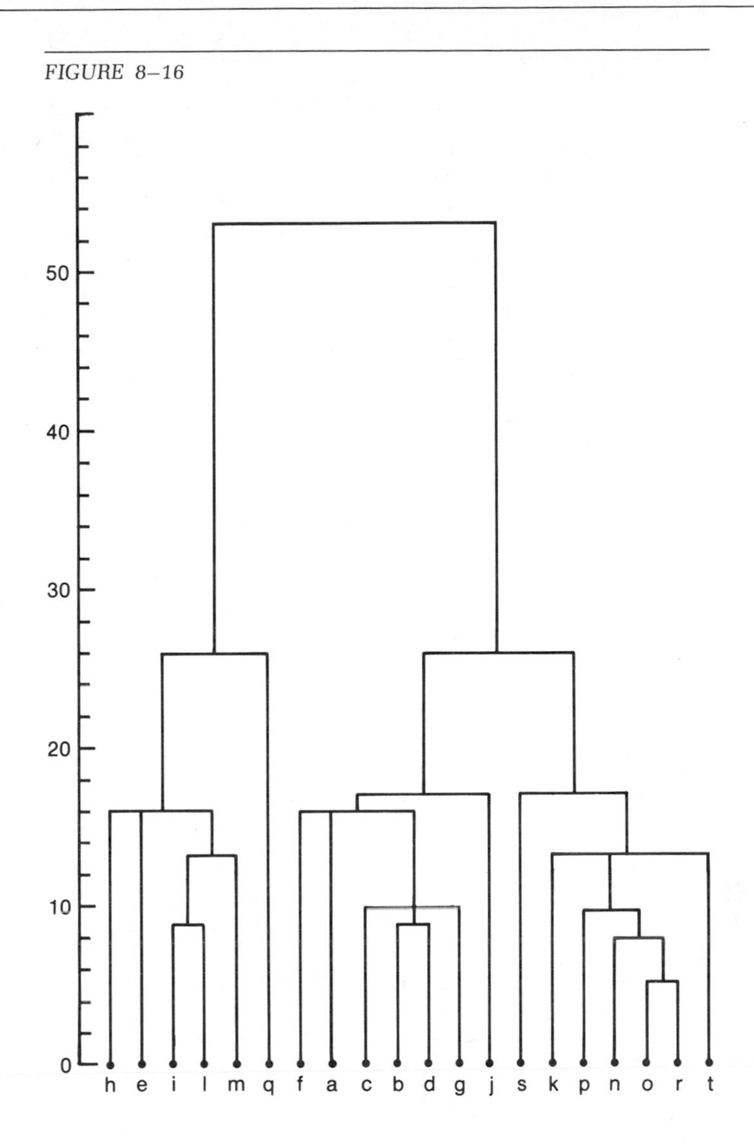

OTHER CLUSTERING PROCEDURES

Large numbers of algorithms exist for performing cluster analysis. There are additional agglomerative and hill-and-valley methods. In addition, there is another class, called divisive methods. Such methods begin with a single cluster and use various criteria to break that cluster down into smaller groups. The methods described above include the most common ones, however, and rather than proceed to consider additional techniques, we turn now to a consideration of distance measures in clustering.

DISTANCE MEASURES

All of the examples used above assumed that we were interested in clustering objects or observations. In addition, they used Euclidean distance or squared Euclidean distance to measure dissimilarity. At this point we return to consider the differences between clustering objects and clustering variables. In addition, we consider other choices for measures of dissimilarity.

As the introductory paragraph to this chapter mentioned, we can cluster either objects or variables. Clustering people according to the amounts they spend for different types of goods is an example of the first type of clustering. Grouping variables in a pilot questionnaire to try to reduce the number of questions is an example of the second. Different measures of dissimilarity are appropriate for the two types of clustering.

With many-valued variables on meaningful scales, such as variables for consumption, the *Euclidean distance* measures are a very common choice. Recall that the squared Euclidean distance between two points is simply the sum of squared differences between the points, dimension by dimension. For example, in three dimensions the squared Euclidean distance between the points (2,5,6) and (6,4,7) is found as follows:

$$(2 - 6)^2 + (5 - 4)^2 + (6 - 7)^2 = 18$$

The Euclidean distance is the square root of this value, or 4.24.

Another measure that is also useful with many-valued variables along a meaningful scale is the *city block,* or *Manhattan metric.* This metric (or measure) defines distances as the sum of the absolute values of differences between points, dimension by dimension. For the two points above, the city block distance is

$$|2 - 6| + |5 - 4| + |6 - 7| = 4 + 1 + 1 = 6$$

The measure emphasizes large differences less heavily than do the Euclidean measures. If, for example, we felt that the difference between \$100 and \$200 in expenditures should be considered twice as important as that between \$50 and \$100 we could use the city-block measure. If we felt, instead, that the larger difference was four times as important, one of the Euclidean measures would be more appropriate.[10]

---

[10]The name city-block, or Manhattan metric, arises from the fact that this measure gives the distance we would have to walk between two points in a city where we could travel only along streets laid out in a square grid pattern (and we could not cut diagonally through any blocks).

In other cases in which we want to group objects, the variables to be used do not give values along a meaningful scale. Such cases are common with variables coded 1 or 0 showing the presence or absence, respectively, of various characteristics. For such cases, there are a range of what are called *matching measures*. For example, we might consider all variables or characteristics on which two objects matched, with both showing 0 or both showing 1. The fraction of all characteristics which matched could then be a measure of similarity. For example, consider the two points in Table 8–14. With this measure the fraction for the two points is 2/6.

Suppose, however, that we were analyzing a situation in which the presence of a characteristic was much more important than its absence. We might be studying the reading habits of individuals and we might ask each person whether s/he had read each of a large list of books. If each person said no to many books, then we would have many 0–0 matches between people. What is really of interest, however, is how many books each pair of people *have* both read. Therefore, we might prefer to count only the 1–1 pairs. Table 8–15 gives such information for several people and then lists the fraction of 1–1 matches for each pair.

Other choices are possible. One possibility is to ignore all 0–0 matches entirely and to define similarity as the 1–1 matches as a fraction of all remaining pairs (the 1–1s, the 1–0s, and the 0–1s). For the first two people in Table 8–15 the measure gives a value of 3/6. For the second and third people in the table it gives 1/11, and so on.

Still other choices are possible. What is important is that investigators select similarity or dissimilarity measures meaningful in the problem under study.

All of the measures described above are for grouping objects. For cases in which we want to group variables, additional measures are needed. In the case of grouping questions in a pilot questionnaire it is natural to find the correlation between answers to different questions and to consider those correlations as a measure of similarity between questions. In fact, a measure based on but different from ordinary correlation is often used. Recall that correlation values lie between −1 and 1. It is common in clustering to use a measure defined as the correlative distance, based on the correlation coefficient but measuring dissimilar-

TABLE 8–14

		Variable					
		1	2	3	4	5	6
Object	*a*	1	0	0	1	1	0
	b	0	1	0	1	0	1

TABLE 8–15

	Variable (book) 1	2	3	4	5	6	7	8	9	10	11	12	13	14
Person a	1	0	0	0	1	1	0	0	0	0	1	0	0	0
b	1	1	1	0	1	0	0	0	0	0	1	0	0	0
c	0	1	0	1	0	1	1	1	0	0	0	1	0	1
d	0	0	1	0	0	1	0	0	0	0	0	0	1	1
e	0	0	1	0	0	0	0	0	1	1	0	0	0	0

Pair	Fraction of 1–1 Matches
ab	3/14
ac	1/14
ad	1/14
ae	0/14
bc	1/14
bd	1/14
be	1/14
cd	2/14
ce	0/14
de	1/14

ity. Suppose that r is the correlation between two variables. The correlative distance for those variables is defined as

$$\sqrt{(1 - r)/2}$$

This quantity is larger for smaller correlation values and vice versa. It has been scaled so that it will take values between 0 and 1.[11]

Correlative distance can be used as a distance measure with variables whose scales make it sensible to consider correlations. For other variables (such as categorical ones), other choices must be found. One possibility is to construct a matching measure for such variables.

SUMMARY

This chapter is essentially a catalog of a few of the choices—for procedures and for distance measures—available to users of cluster analysis. The approach seems apt because cluster analysis is really a large collection of techniques for grouping. Some of those techniques are better in some situations than are others. As noted above, the nearest-neighbor

[11]The precise definition of correlative distance is used because it turns out to be analogous to Euclidean distance if it is applied to objects, rather than variables. In fact, if objects (not variables) are standardized and plotted, the correlative distance is half the ordinary Euclidean distance between points.

method will chain together points that are not compact (as in Figure 8–13), while the farthest-neighbor and minimum-squared-error methods are good for finding compact, homogeneous groupings. The hill-and-valley methods are appropriate when investigators believe that the concept of density is useful in their problems and that they want to use a method that focuses on the most dense points.

In part, investigators' choices of methods depend on which of the wide variety of clustering procedures are available on their computers. It is essential, however, that users also consider what methods are appropriate for their specific problems. Anderberg's playing-card example, given at the start of this discussion, emphasizes what is the critical point of cluster analysis: Different groupings of the same set of entities are most meaningful under different circumstances. Cluster analysis is situation-specific. The choice of technique depends on the context and the purpose of the analysis.

APPENDIX: TO FACTOR OR TO CLUSTER

As described in the preceding two chapters, factor analysis and cluster analysis are distinct techniques (or sets of techniques) for data analysis. The literature shows considerable confusion about the similarities and differences between factor analysis and cluster analysis and also about the types of situations in which each of the methodologies is appropriate. The literature also contains examples of the misuse of the methods—particularly, the use of factor analysis in cases for which some form of cluster analysis would be the appropriate choice.[12]

Part of the confusion between the techniques likely arises because both methodologies can be applied either to observations (objects) or to variables. Thus, the possible situations form a two by two matrix as shown in Figure 8A–1. Other issues complicate the situation. For example, investigators must choose one of many methods for cluster analysis. Figure 8A–1, however, captures the major cases that should be distinguished.

This appendix reviews briefly the basic objectives and underlying assumptions of factor analysis and cluster analysis, with the aim of distinguishing the methodologies from one another. It then proceeds to

[12]For example, in his extensive review article on market segmentation, Yoram Wind (1978) reports that factor analysis is often used by investigators in marketing to reduce the numbers of questions included in questionnaires. This discussion takes issue with that reported practice—although not, of course, with Wind's accurate reporting of what the practice is. Cluster analysis rather than factor analysis would be the appropriate choice. (Yoram Wind, "Issues and Advances in Segmentation Research," *Journal of Marketing Research*, August 1978, pp. 317–37).

FIGURE 8A–1

	Observations (objects)	Variables
Factor Analysis		
Cluster Analysis		

consider, in turn, each of the cells in the two by two matrix. Finally, the discussion ends by considering some sets of data that could be analyzed in more than one of the four cells.

FACTOR ANALYSIS VERSUS CLUSTER ANALYSIS

The basic difference between factor analysis and cluster analysis is the following: Factor analysis helps to dissect or to examine the internal structures of entities (objects or variables). Cluster analysis groups similar entities (objects or variables), taken as wholes.

Factor analysis of variables for final grades of high school students might dissect those grades into parts depending on writing ability, logical thinking, speaking ability, and so on. Factor analysis of variables giving viewers' ratings of TV shows might dissect the ratings into parts depending on interest in comedy, interest in music, interest in news, and so on. In each case, investigators use factor analysis to help identify underlying unobservable characteristics (factors or dimensions) *within* entities. Before the analysis, the investigators do not know how to describe the internal structures of those entities.

Cluster analysis might identify groups of minicomputers that are similar to one another. Or, investigators might cluster individuals into market segments on the basis of their spending habits. Clustering techniques group entities which are similar in some *overall* sense. At the outset, investigators must specify how similarity is to be measured. Before the analysis, they do not, however, know what groupings exist or are meaningful.

R-TYPE FACTOR ANALYSIS

The most common type of factor analysis, *R*-type, applies to variables. The text chapter on factor analysis describes *R*-type factor analysis. The classic example is the analysis of examination grades; Spearman used that example in devising factor analysis in the early 20th century.

We use such factor analysis in situations in which we have a set of variables that are correlated and for which we want to try to explain the connections among the variables—to try to identify and understand the linkages (or factors). It was for this purpose that Spearman devised factor analysis. In using factor analysis, investigators do not know at the start how many factors or which factors underlie the connections among a set of manifest (or observable) variables. (It is usually for precisely this reason that they are using the technique.)

The two-factor model for students' grades that was described in the chapter on factor analysis was:

$$Z_j = a_{j1}f_1 + a_{j2}f_2 + m_j + d_j u_j$$

Here Z_j is the variable for grade on the jth test, f_1 and f_2 are the two common factors, a_{j1} and a_{j2} are the loadings, m_j is the mean of the grades on the jth test, u_j is ability specific to the jth test, and d_j is a constant specific to the jth test. Assume that we have decided that f_1 measures quantitative ability and f_2 measures verbal ability.

This basic factor analytic model involves very specific assumptions about reality. It assumes that each student (observation) possesses a level (or amount) of verbal ability (f_2) and a level of quantitative ability (f_1). Each student also possesses levels of each of the test-specific abilities $u_1, u_2, \ldots, u_6$. The factor analysis model assumes a relatively simple linear world. Each grade ($Z_1, Z_2, \ldots, Z_6$) is a simple linear function of the two common factors (f_1 and f_2) and the test-specific ability for that examination. Thus, for example, there are not interactions between the factors or between one factor and a test-specific u_j. Further, the model assumes that all correlation among the examination grades occurs (or is explained) through the common factors f_1 and f_2. Different tests can have different communalities.

In general, as in this example, the R-type factor analytic model assumes that unobservable factors are *possessed by observations*. (Here the abilities are possessed by students.) The factors explain to varying degrees the values on those observations of each of a group of variables (the tests). More precisely, the factors explain the correlations among the values of the variables.

Normally in applications of factor analysis investigators do not know how many factors or which factors underlie a set of manifest variables. Instead, the aims of the analysis are to determine the number of factors, to label or identify the factors, and in many cases, to estimate the extent to which each observation possesses each factor. For some specified possible number of factors, a computer program for factor analysis produces a set of estimates of factor loadings—the a_{jp} from the factor analytic model. Investigators use their prior knowledge about the variables to try to label the factors on the basis of the patterns of loadings. For example, an investigator in the examination-grades problem might have noticed that the mathematical grade loaded heavily on the first factor, that the chemistry grade had a medium loading for that factor, and that the other variables had small loadings on that factor, with Latin a bit higher than the other three. The investigator's knowledge of the six examination subjects might then have suggested that the first factor could be interpreted as quantitative ability.

Once an investigator has determined that the factor analysis results with some specific number of factors are meaningful, the computer can estimate the values of each factor (factor scores) possessed by each observation. In the grades example, the computer would estimate each student's verbal ability score and quantitative ability score. In the process, the program could create new variables (the two factor scores) that could be used in further analyses. The verbal and quantitative scores, for example, might be used to predict individual students' grades on some seventh test.

Such *R*-type factor analysis could sensibly be applied to a number of problems in marketing and other fields. For example, the technique has been applied to subjects' preference ratings of TV shows, with each subject as an observation and the ratings of the shows as variables, one variable for each show. The use of *R*-type factor analysis involves the assumption that an individual's ratings of the shows are tied together through underlying unobservable factors. By examining the loadings of various shows on a factor, we would try to interpret that factor. The first might tentatively be identified as interest in comedy shows, the second as interest in educational or documentary programs, the third as interest in music, and so on. Each observation would be assumed to possess each of these factors to some degree. One person might have a high score for interest in comedies and a high score for interest in documentaries, another might have high interest in comedies but low interest in documentaries, and so on.

The loadings would link the preference ratings (variables) and the factors. It would seem likely, for example, that the rating for "Mork and Mindy" would load heavily on the factor for interest in comedy. Similarly, the rating for "60 Minutes" would likely load heavily on the documentary-interest factor. The rating for "Saturday Night Live" would likely load heavily on comedy and music. In addition, there would be a show-specific part for each show. (Each person would have a value for each show-specific part.) The show ratings would depend to varying degrees on the common factors and on the show-specific values.

The aims of using factor analysis in this situation would include those of decomposing the TV show ratings (to determine which ratings depended strongly on which interest factors) and those of estimating the extent to which each person possessed each factor. The estimated factor scores might be used, with other analytic techniques, to group the observations (people) into market segments.

Another possible application of *R*-type factor analysis would involve subjects' preference ratings of a group of products. For example, we might consider cars. The results of such an analysis might identify three basic factors possessed by the respondents: style-consciousness, price-consciousness, and performance-consciousness. The loadings estimated by the factor analysis procedure would show the extent to which ratings for different cars depended on individuals' possession of the underlying factors. For example, the rating of the Mercedes sedan would likely load heavily on style-consciousness and on performance-consciousness.

In summary, *R*-type factor analysis is used to study underlying factors through which a set of manifest variables are linked. The model is a linear one. The factors, as well as variable-specific values (u_js), are possessed by the observations. Each manifest variable is decomposed through factor analysis into the sum of parts depending on the various underlying factors plus a part specific to that manifest variable.

Q-TYPE FACTOR ANALYSIS

In *Q*-type factor analysis, the roles of the variables and the observations are switched from what they are in *R*-type. Underlying common factors are possessed by variables and are used to decompose observations. This type of analysis is factor analysis applied to observations.

As an example, we might want to consider factors of sports footwear which determine people's preferences for such shoes. As data we could collect several individuals' preference ratings for a variety of models of sports shoes. The algebraic formulation would involve one equation for each person (rather than one for each variable, as in *R*-type analysis). Suppose P_i denotes person *i*'s preference rating; P_i will take on a value for each of the shoe models. The three-factor model would be

$$P_1 = \alpha_{11}f_1 + \alpha_{12}f_2 + \alpha_{13}f_3 + \delta_1 u_1 + m_1$$

and so on, for as many equations as there are people. In general,

$$P_i = \alpha_{i1}f_1 + \alpha_{i2}f_2 + \alpha_{i3}f_3 + \delta_i u_i + m_i$$

In this equation, α_{ij} is the loading of the ith person on the jth factor. u_i is a variable specific to the ith person. m_i is the mean rating given by the ith person.

Once we had the output of a three-factor analysis of this problem, we would examine the set of loadings of the different individuals on each particular factor to try to interpret that factor. In doing so, we would be using our prior knowledge about the individuals. (The situation is analogous to the interpretation step in *R*-type analysis, where we would use the factor loadings and our prior knowledge of the examination subjects to interpret the factors.) For example, if the people with high loadings on the first factor were all strongly competitive runners we might interpret that factor as performance.

In Q-type analysis, the factors are qualities of the variables (the shoe products). Suppose that after analyzing the results of such an analysis we decided that there were in fact three underlying factors. We might, for example, interpret the three factors as measuring performance, style, and price. In that case, the interpretation means we believe that the shoe models are characterized by performance, style, and price. An individual's rating of a particular shoe model would depend on the extent to which the shoe possessed each of the features and on a person-specific variable. Here f_j is the value of the jth factor or feature for the shoe model being considered. α_{ij} is the extent to which individual i emphasizes the jth feature. u_i is the value specific to person i. ${\alpha_{i1}}^2 + {\alpha_{i2}}^2 + {\alpha_{i3}}^2$ is the communality of person i's ratings, the portion of the variance of the values of P_i that is explained by the common factors. u_i might be interpreted as the idiosyncratic part of person i's preferences. (Recall that u_i is assumed independent of u_k in the factor analytic model.) ${\delta_i}^2$ is the portion of the variance of P_i that is attributable to this idiosyncratic part.

For Q-type factor analysis to make sense we must be willing to assume that the individuals under study perceive the variables according to the same set of common features (factors), plus one idiosyncratic variable each. Individuals must agree on the extent to which each variable (product) possesses each factor. Each person's preference ratings must be a linear function of the factors possessed by the shoes. The coefficients in the linear functions for different people will in general differ, however. In other words, individuals will generally assign different importances to the various factors. (α_{11} will be different from α_{21}.) The factors are possessed by variables. They are used to decompose or look inside the observations (the preference-rating mechanisms of individuals).

To contrast Q-type and R-type factor analysis we can use the R-type analysis of cars and the Q-type analysis of sports shoes. In the R-type analysis, people had levels of each factor. There was one equation per car, decomposing preferences for the car into portions due to each common factor plus a car-specific part. Factors were such things as style-consciousness. To interpret or name factors such as this, investigators would examine the full set of loadings for the factor (one from each equation). By noticing the magnitudes and signs of the loadings and by considering their own knowledge of the car models to which those loadings corresponded, investigators could try to label the factor.

In the Q-type analysis, by contrast, the shoe models possessed levels of each factor, which were things like style. There was one equation per person. The idea that factors belonged to shoes might make this type of analysis attractive. On the other hand, the interpretation or identification of the factors can be very difficult in Q-type analysis. As noted above, investigators would use the loadings for a particular factor (with one loading per person), together with their knowledge of the *people* involved, to guess at a label or interpretation for the factor. Rarely do investigators have such knowledge of their observations (as opposed to the variables in their problems).

CLUSTER ANALYSIS OF OBJECTS

Cluster analysis deals with similarities between entities viewed as wholes. The most common use is in the grouping of objects (or observations). For example, one of the problems for which the techniques of cluster analysis were devised is the classification of organisms in biology (where the techniques of cluster analysis are called numerical taxonomy). To cluster a number of organisms we might record the presence (1) or absence (0) for each organism of each of a large set of features. One such feature could be fins, another webbed feet, and so on. There would be one dummy variable per feature and one observation per organism.

Before running a cluster analysis, we would have to specify a measure of similarity or dissimilarity (distance). We might, for example, assume that a feature was important in comparing two organisms only if at least one of those organisms possessed the feature. In other words, we might completely ignore all 0–0 matches (such as the fact that cows and pigs both lack fins). If so, we might define as a matching measure of similarity between two organisms the number of 1–1 matches (features possessed by both) divided by the sum of the numbers of 1–1, 1–0, and 0–1 matches (features possessed by at least one). Of course, many other measures of similarity or dissimilarity are possible.

We would also have to select an algorithm for grouping. We could then apply the chosen technique and the chosen distance measure to the data. Notice that we *started* by specifying the variables (features) to be considered, together with the way for computing distances on the basis of those variables. The procedure then grouped observations. No decomposition of either variables or observations was involved. The contrast with factor analysis should be clear.

Clustering of observations can be applied to many problems in marketing and related fields. For example, for each of a group of respondents we might record whether (1) or not (0) each person owned each of a large number of products. There would be one variable per product and one observation per respondent. We would have to select a measure of similarity or dissimilarity of ownership. We might, for example, use as a similarity measure between two people the number of products which both owned (1–1 matches) divided by the total number of products (1–1, 1–0, 0–1, or 0–0). We could then use a clustering procedure to group people and could try to interpret (and use) the resulting groups, at some stage in the grouping process, as market segments.

Similarly, we might use cluster analysis to study the market for a set of industrial products such as machine tools. For each product (observation) we would record values for each of a set of objective characteristics (variables). Price could be one such characteristic. Various product features could be others. We would have to specify a distance measure based on those characteristics and also select a clustering algorithm. We could try to use the resultant groupings to understand product positions in the marketplace.

Cluster analysis could also be used with preference ratings. For example, each of a group of purchasers could be asked for a preference rating for each of several vendors of industrial supplies (such as lubricants and coated abrasives). There would be one preference rating (variable) per vendor and one observation per purchaser. Thus, for each purchaser we would have a vector of ratings. We could try to use those vectors to group the purchasers into market segments. We would select a distance measure (such as correlative distance) and an algorithm, and we would try to interpret the results in terms of market segments.

In all of these examples, observations are clustered as wholes. The values of variables on those observations are used in calculating user-specified distance measures. The algorithms do no decomposition—of either variables or observations.

CLUSTER ANALYSIS OF VARIABLES

In cluster analysis of variables we also group entities as wholes, but in this case we group variables rather than observations. Such a procedure might be used, for example, as one step in the refinement of a questionnaire for use in a marketing research study.

Suppose that a pilot questionnaire with a large number of questions had been devised and administered to a sample of respondents. Suppose further that it was impractical to use that many questions on the final questionnaire, so that one purpose in analyzing the pilot results was to eliminate some questions. It would seem sensible to try to eliminate questions that overlapped other questions. In other words, if a question elicited information that was also elicited by other questions, we would lose little by eliminating the first question. Accordingly, we might try to find groups of questions with highly correlated response patterns. We could use such groupings as the basis for eliminating questions.

Each question's responses would be a variable in this analysis. Each respondent would be an observation. The correlation coefficient r between pairs of variables could be used as the basis of a distance measure. We might use the correlative distance (defined as $\sqrt{.5(1 - r)}$). We could select a clustering procedure designed to form tightly connected clusters. (The farthest-neighbor procedure would be a good choice.) We could then eliminate from the questionnaire all but one or two of the questions in any one of the tightly connected clusters.

This example is an appropriate one for demonstrating the confusion that seems to exist between cluster analysis and factor analysis. As noted by Yoram Wind, investigators sometimes apply factor analysis to this problem in trying to find a basis for eliminating questions. Such use is a misapplication of factor analysis. The aim is not to look inside or decompose either questions (variables) or respondents (observations). Rather, it is to group similar questions, taken as wholes. Cluster analysis is the appropriate technique.

On the other hand, there would be a sensible use for factor analysis of the pilot questionnaire if our purpose and plans for analysis were slightly different. Suppose we did want to look inside the variables (the questions) to determine what basic dimensions the questionnaire was measuring. We might run factor analysis and then, after interpreting the resultant factor loadings, select new questions that more clearly and cleanly addressed the basic dimensions we had uncovered. In this process, our aim would not be to group the current questions, as wholes. It would instead be to dissect those questions so as to identify and then address directly their underlying factors. In that case, factor analysis would be an appropriate technique to use.

CHOICE OF METHOD

To conclude this discussion, it is useful to consider situations in which investigators could sensibly make different choices of techniques, depending on their purposes.

First, consider problems of market segmentation. Either cluster analysis or factor analysis could be used. The discussion above of cluster analysis of observations described how manifest variables coded for product ownership (1) or not (0) could be used to group respondents. That section also described how cluster analysis could be applied to the preference ratings assigned by industrial purchasers to vendors; the analysis would group the purchasers. In either of these examples, respondents are being grouped, as wholes.

In the section on *R*-type factor analysis, on the other hand, one example considered how the technique could be used to identify factors such as style-consciousness. The level of each such factor possessed by each respondent (the factor scores) could then be estimated. Here the emphasis is on looking within respondents to identify the levels of factors possessed by those respondents. Once factor analysis had been used to identify factors and estimate factor scores, respondents might be grouped into market segments on the basis of those scores. Cluster analysis would be appropriate for performing the grouping.

Q-type factor analysis could also be used in segmentation studies. The section on that technique described a problem involving factors of performance, style, and price possessed by sports footwear, with one equation per respondent. The loadings in such an equation reflected the importance levels attached by the respondent to performance, to price, and to style; the equation also showed the importance of the respondent-idiosyncratic portion. After factor analysis had been used to estimate the loadings, respondents with similar patterns of loadings (levels of importance) could be grouped into market segments. Cluster analysis could perform the grouping.

Sometimes factor analysis or cluster analysis could usefully be applied to exactly the same set of data, depending on the purpose of the analysis. The questionnaire example in the previous section is one such case. For another example, suppose we had observations giving respondents' preference ratings for TV shows. There would be one variable per show. *R*-type factor analysis could be used to decompose the ratings into parts, such as those dependent on interest in music or on interest in comedy. In this application, the emphasis would be on looking within the ratings to see the importance of the various factors in the ratings.

On the other hand, we could use cluster analysis with the same data. The ratings could be converted to measures of distance between shows. The idea would be that similar shows would be viewed or rated similarly (either positively or negatively) by respondents. The distance measures could then be used in a clustering procedure for grouping shows (variables). Here the emphasis is on the overall similarity of the shows viewed as wholes.

In summary, both cluster analysis and factor analysis can be applied either to variables or to observations. There is unfortunate confusion over which technique is appropriate for particular problems. There are examples of misuse, especially the use of factor analysis where cluster analysis should be chosen. The basic difference between the techniques is not in the type of data they require, since, as shown, they can sometimes be applied to exactly the same sets of data. Instead, factor analysis and cluster analysis serve different analytic or investigatory purposes. With factor analysis investigators can look inside or decompose entities (variables or observations). With cluster analysis they group entities which, overall, are similar to one another.

CHAPTER 9

Multidimensional Scaling

Multidimensional scaling and related techniques are used to explore the information contained in subjects' assessments of the distances (or dissimilarities) among a collection of objects. For example, the techniques could be used on assessments by consumers regarding the differences among various models of passenger cars. In general, the techniques convert distance or dissimilarity information into geometric representations of the objects in some number (or several different numbers) of dimensions. Often, investigators want to interpret the resulting dimensions. In other cases, the objective is to use the geometric placements of the objects to study groupings or clusters of those objects.

Basic to the scaling techniques is the assumption that underlying the perceptions by a subject of various objects (such as car models) are some number of attributes or dimensions. Further, it is assumed that the dissimilarities perceived by the subject among the objects are determined, except perhaps for random errors, by the geometric distances between the objects in the appropriate number of dimensions. For example, with two dimensions and five objects (A, B, C, D, and E) suppose that the objects were located relative to some subject's actual underlying dimensions as shown in Figure 9–1. The assumptions of the scaling techniques would then imply that the subject believes:

1. Objects A and B are the most similar of all possible pairs of the five objects—because the distance between the corresponding points is smallest.
2. That E is more like C than it is like D—because the distance from E to C is less than that from E to D, and so on.

In general, even if we believe that subjects do perceive dissimilarities according to distances between objects when those objects have

FIGURE 9–1

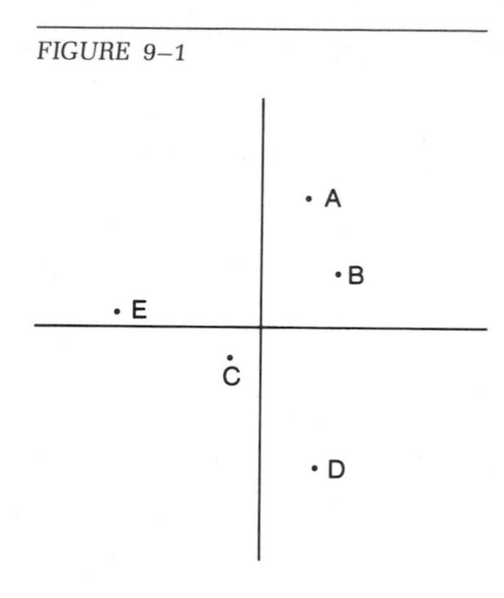

been located with reference to underlying attribute dimensions, often we cannot ask subjects directly to tell us what are the dimensions and what are the coordinates of each object. Sometimes we feel that subjects cannot readily identify the dimensions. In other cases, we may feel that subjects could identify dimensions but that they would not be able to specify the coordinates of the objects along those dimensions. In many cases, we feel it more sensible to ask subjects simply to make judgments about objects taken as wholes. In other words, we do not ask subjects to deal explicitly with underlying dimensions. Instead, we ask for judgments about dissimilarities among objects considered as wholes and we use scaling techniques to extract information about the underlying dimensions and about the relative locations of the objects within the space defined by those dimensions.

The precise form of the required judgments varies. In some situations we ask subjects for quantitative assessments of the dissimilarities between pairs of objects. For example, we may assume that subjects can give us their perceptions of the actual distances between objects. In such cases, we say that the subjects provide *metric* inputs (actual distance measures); we assume that the input variables have full cardinal properties. In other situations, we assume instead that subjects can give us only rank orders of the dissimilarities. In other words, we assume that the subjects can tell us which pair of objects is most similar, which pair is next most similar, and so on, but we do not assume that they can expand the information in the ordering of pairs to give actual distances. The judgments in such cases are ordinal variables. To distinguish them from metric input judgments, we also call them *nonmetric* inputs.

EXAMPLES OF APPLICATIONS OF SCALING

To have examples for the explanations that follow, we will now consider two problems to which scaling methods can be applied. The first is taken from a discussion by J. B. Kruskal and M. Wish of Bell Labs.[1] It involves airline distances rather than preference information and is intended to show how scaling algorithms can extract geometric information from true distance data. The inputs for this first example are given in Table 9–1.

The second example is a far more typical application of scaling ideas. A subject who is knowledgeable about the computer industry was asked to make judgments about the similarities and dissimilarities among 10 computer companies. The questions did not concern specific computer models but dealt with the firms. The idea was to determine the subject's perceptions of the brand images of the various companies. For this example, only rank order information was used. The subject was presented with 45 cards, each labeled with the names of a pair of the 10 firms.[2] After considerable work, the subject put those cards into the order summarized in Table 9–2. In that table, the most similar pair is labeled 1 and the least similar pair is labeled 45.

The reader will note that Table 9–2 lists each pair of companies only once. It lists only Amdahl—Itel, not Amdahl—Itel and Itel—

TABLE 9–1
Airline Distances between Ten U.S. Cities

	Atlanta	Chicago	Denver	Houston	Los Angeles	Miami	New York	San Francisco	Seattle	Washington, D.C.
Atlanta										
Chicago	587									
Denver	1212	920								
Houston	701	940	879							
Los Angeles	1936	1745	831	1374						
Miami	604	1188	1726	968	2339					
New York	748	713	1631	1420	2451	1092				
San Francisco	2139	1858	949	1645	347	2594	2571			
Seattle	2182	1737	1021	1891	959	2734	2408	678		
Washington, D.C.	543	597	1494	1220	2300	923	205	2442	2329	

[1] Joseph B. Kruskal and Myron Wish, *Multidimensional Scaling* (Beverly Hills: Sage Publications, 1978).

[2] There are exactly 45 different pairs that can be formed from 10 objects if we assume that order does not matter.

TABLE 9–2

1.	Amdahl—Itel	24.	DEC—Amdahl
2.	DEC—Hewlett Packard	25.	DEC—Itel
3.	DEC—Data General	26.	Burroughs—Itel
4.	Data General—Hewlett Packard	27.	Control Data—Cray
5.	IBM—Burroughs	28.	Amdahl—Cray
6.	IBM—Control Data	29.	Hewlett Packard—Burroughs
7.	IBM—DEC	30.	Burroughs—Cray
8.	Control Data—Burroughs	31.	Hewlett Packard—Amdahl
9.	Data General—Rolm	32.	Data General—Amdahl
10.	IBM—Hewlett Packard	33.	Data General—Itel
11.	IBM—Data General	34.	Data General—Burroughs
12.	DEC—Rolm	35.	Hewlett Packard—Itel
13.	IBM—Amdahl	36.	Hewlett Packard—Cray
14.	IBM—Itel	37.	IBM—Rolm
15.	Control Data—Amdahl	38.	Burroughs—Rolm
16.	Control Data—Itel	39.	Amdahl—Rolm
17.	Burroughs—Amdahl	40.	Itel—Rolm
18.	DEC—Control Data	41.	IBM—Cray
19.	DEC—Burroughs	42.	DEC—Cray
20.	Hewlett Packard—Control Data	43.	Itel—Cray
21.	Data General—Control Data	44.	Data General—Cray
22.	Control Data—Rolm	45.	Cray—Rolm
23.	Hewlett Packard—Rolm		

Amdahl. The (sensible) assumption is that the pair has only one dissimilarity value, regardless of which listing of the two names is used. Table 9–2 includes the pairs from the triangular array in Figure 9–2. In the listings, the name on the row is given first.

One of the tasks for the analyst in using scaling is to specify what measure of distance is to be used. For the time being, we will consider only ordinary Euclidean distances.[3] Alternative choices will be considered later in the discussion. In all cases, investigators should use the distance measure that they believe their subjects use in their perceptions of dissimilarities.

METRIC SCALING

Procedures for *metric scaling* treat the input data as exact distances between pairs of objects. Such procedures make no attempt to determine whether or not the inputs are in fact exact distances. They simply

---

[3]The Euclidean distance between two points is defined as follows: Suppose that we are dealing with m dimensions. Find the difference between the coordinates of the two points on the first dimension and square it. Similarly, find the squared difference of the coordinates on each other dimension. Add these m squared differences. The Euclidean distance is the square root of this sum. Euclidean distance is the standard measure used routinely in measuring distances.

FIGURE 9–2

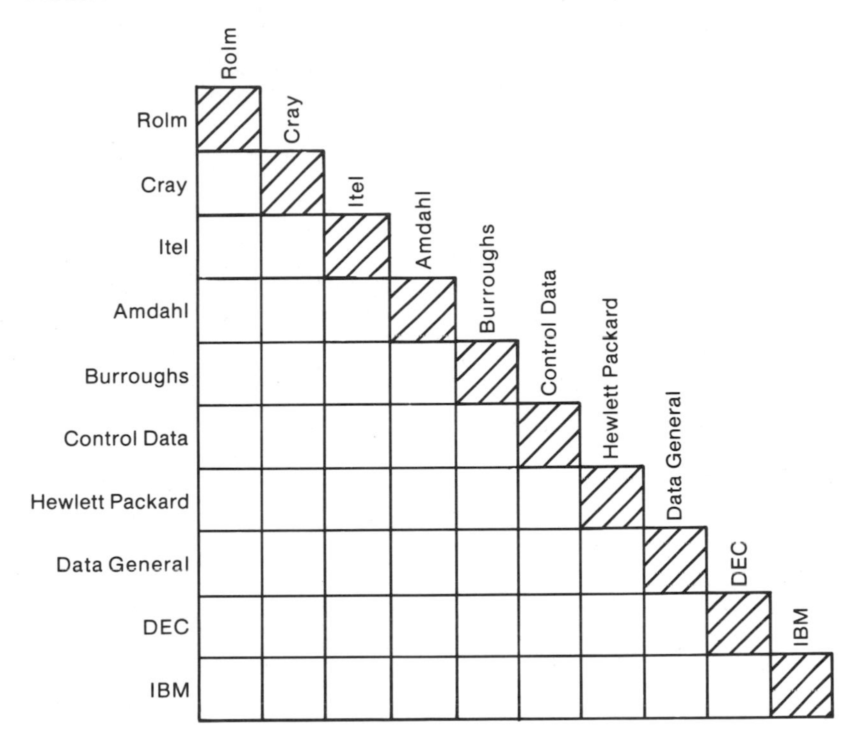

construct geometric representations of the points that correspond as closely as possible with the input "distances." Thus, the basic idea is to convert metric (distance) inputs into geometric output.

For the data on airline distances between cities, a metric scaling algorithm produced the map in Figure 9–3. The program was given the inter-city distances and was instructed to use two dimensions. The map is very close to a map of the United States (rotated so that west is at the lower left and east is at the upper right).

In this simple example, it was obvious that we would want to consider two dimensions in plotting the locations of cities. In most scaling problems, the dimensionality is not obvious at all. For example, suppose we are considering distances between perceptions of pairs of cars. (Assume for now that the subject has in fact been able to give us distance values.) It will generally not be clear how many dimensions to use. Computer programs for scaling require users to specify how many dimensions to consider; the programs do not select the dimensionality. If in fact the subject perceives the objects according to exactly m underlying dimensions and if in addition the subject has given the true interpoint distances, then for metric scaling the following will be true:

FIGURE 9–3

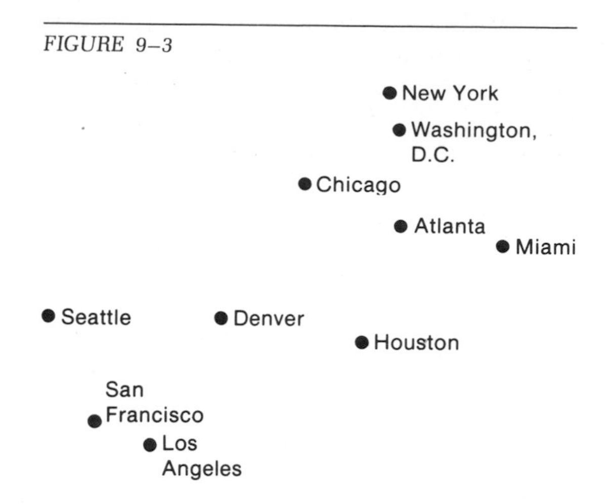

1. If we tell the computer program to use m dimensions, it will construct the true configuration of the points.
2. If we tell the computer to use some number of dimensions m^*, with m^* larger than m, the program will find the true configuration in the first m coordinates and will set the remaining $(m^* - m)$ coordinates equal to zero for all of the points.
3. If we tell the program to use some number m^* of dimensions where m^* is smaller than m, the program will find the first m^* principal components (principal axes) of the true configuration.[4]

Metric scaling has been introduced here with an example for which the assumptions are very reasonable (although the issue of why one would actually be performing scaling is considerably less clear). The reader is likely wondering whether the metric scaling procedures ever actually make sense for real problems, in which it is rather unlikely that subjects can give exact distance measures. Even in situations where subjects feel they can specify distances, they are likely to provide distances plus errors, at best. In many other situations, subjects will not feel capable of giving either distances or near-distances. Providing rank orders will be a sufficient challenge for them.

There are three basic ways to approach these problems. First, rather surprisingly, it turns out that for many problems metric scaling works very well, even if we have only approximate distances or, in some cases, only rank orders. This fact is an empirical one rather than one based on the theory of scaling. Investigators state that metric scaling is

[4]See Chapter 6 on principal components analysis.

amazingly robust, or able to perform well with data which do not satisfy the theoretical requirements for input data.

The second approach is to use what is called quasi-nonmetric scaling. Such scaling begins with an application of a metric scaling algorithm. Rather than stopping and outputting the resultant configuration of points, in quasi-nonmetric scaling the computer attempts to modify distance values slightly so that they can be fit more closely as true distances. The modified input values can be called enhanced distances. The computer applies the metric algorithm to the enhanced distances. It may then repeat the cycle several times, further enhancing or improving the distance values and then applying the metric algorithm to find a configuration. This chapter will not describe quasi-nonmetric scaling in further detail. Instead, we turn to nonmetric scaling, which is the third approach for handling input data which do not give exact distances.

NONMETRIC SCALING

In nonmetric multidimensional scaling the computer program takes as inputs a subject's rank ordering of the distances or dissimilarities between pairs of objects.[5] The procedure assumes that the inputs are not actual distances (or even actual distances plus errors) but that they do contain ordinal information. The program also requires instruction on what number of dimensions to consider. It then proceeds to attempt to find the configuration of objects in the specified number of dimensions which gives distances that best fit the input rank orders. It is necessary to consider a number of details of this procedure. First, however, it is useful to examine one set of outputs from a nonmetric scaling program and, in the context of that example, to discuss the major purposes of the analysis.

The following figures summarize a portion of the output obtained from program KYST, perhaps the best-known scaling program, when it was applied to the dissimilarities judgments on computer companies in Table 9–2. Two sets of results are given. Figure 9–4 shows the three graphs that were obtained when the program was told to consider three dimensions. There is one graph for each pair of dimensions. Figure 9–5 shows the single graph obtained when KYST was told to consider only two dimensions. In each case, KYST provided several other types of output, but we defer the discussion of those additional outputs.

In examining output from KYST, we generally have several concerns. First is interpretation. We would like to determine which objects

[5]For the computer companies, the program inputs are the rankings in Table 9–2.

FIGURE 9–4
Scaling in Three Dimensions (dimensions 1 and 2)

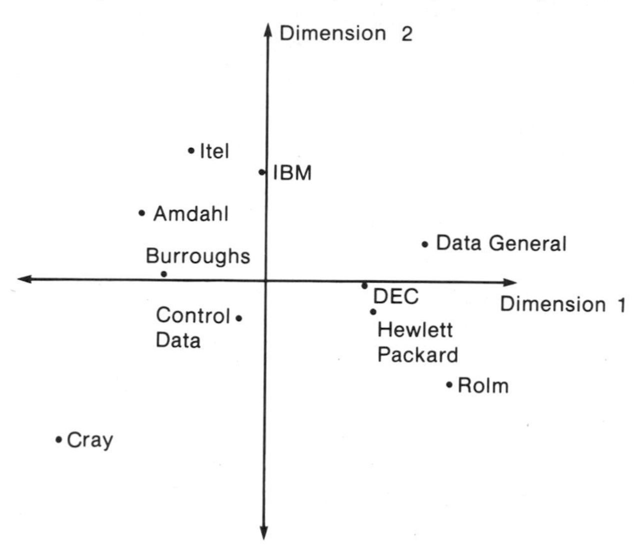

FIGURE 9–4 *(continued)*
Dimensions 1 and 3

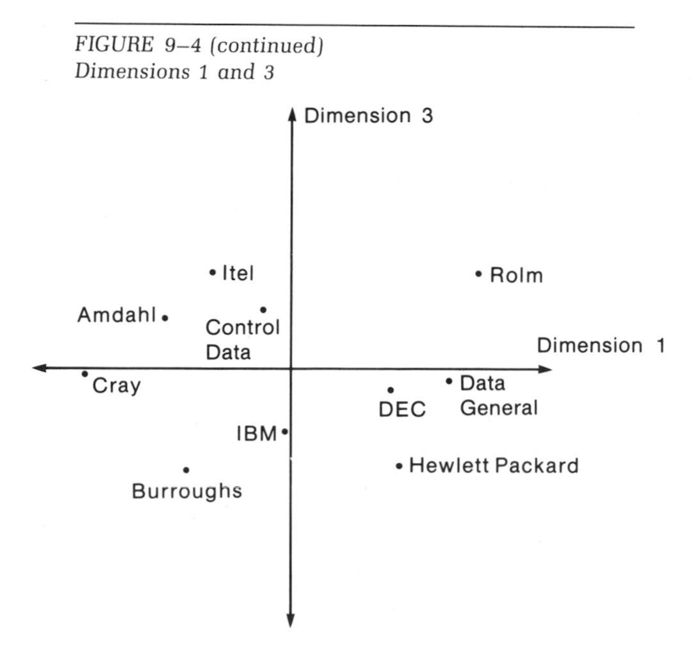

FIGURE 9–4 (concluded)
Dimensions 2 and 3

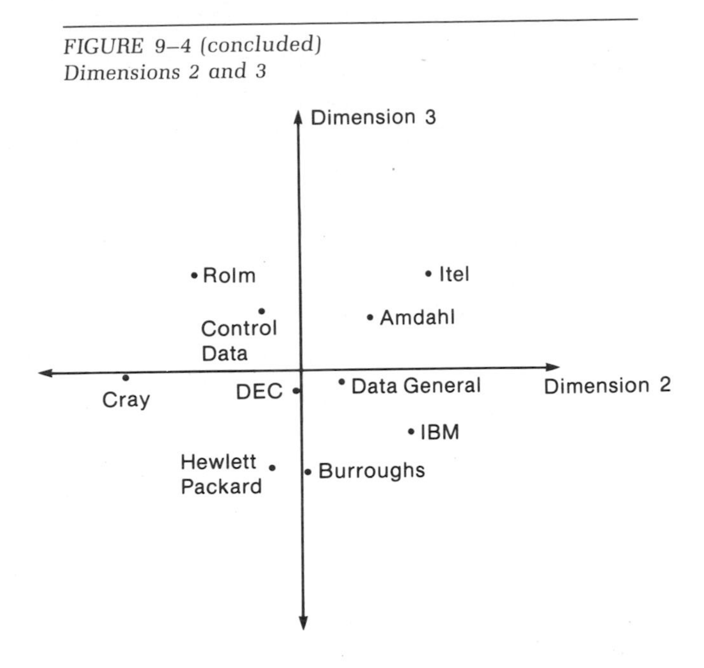

are like each other and which are unlike. We would like to see which groups of objects cluster together.[6] Often, we would also like to interpret the axes in the figures, determining which attributes the subjects seem to be using in comparing objects. Other concerns arise. The output given here considers only two possible choices for the number of dimensions. Clearly the computer could have been instructed to consider other possibilities as well. We face the problem of deciding which number of dimensions to use in describing the subject's perceptions. In addition, we will want somehow to measure how well the computer-fit configuration matches the subject's judgments. In other words, we ask how closely the distances from the configuration actually fit the input dissimilarities.

As would be expected, the answers to these concerns are interrelated. In general, we try several different dimensionalities in using scaling. For each, we consider interpretability and we also consider goodness-of-fit (as defined more precisely below). More dimensions will in general allow better fit but will also generally be harder to interpret. Hence, our final choice of dimension is often a compromise.

[6]We might want to subject the results of multidimensional scaling to formal cluster analysis to group similar objects or simply to examine the output of scaling judgmentally.

FIGURE 9–5
Scaling in Two Dimensions

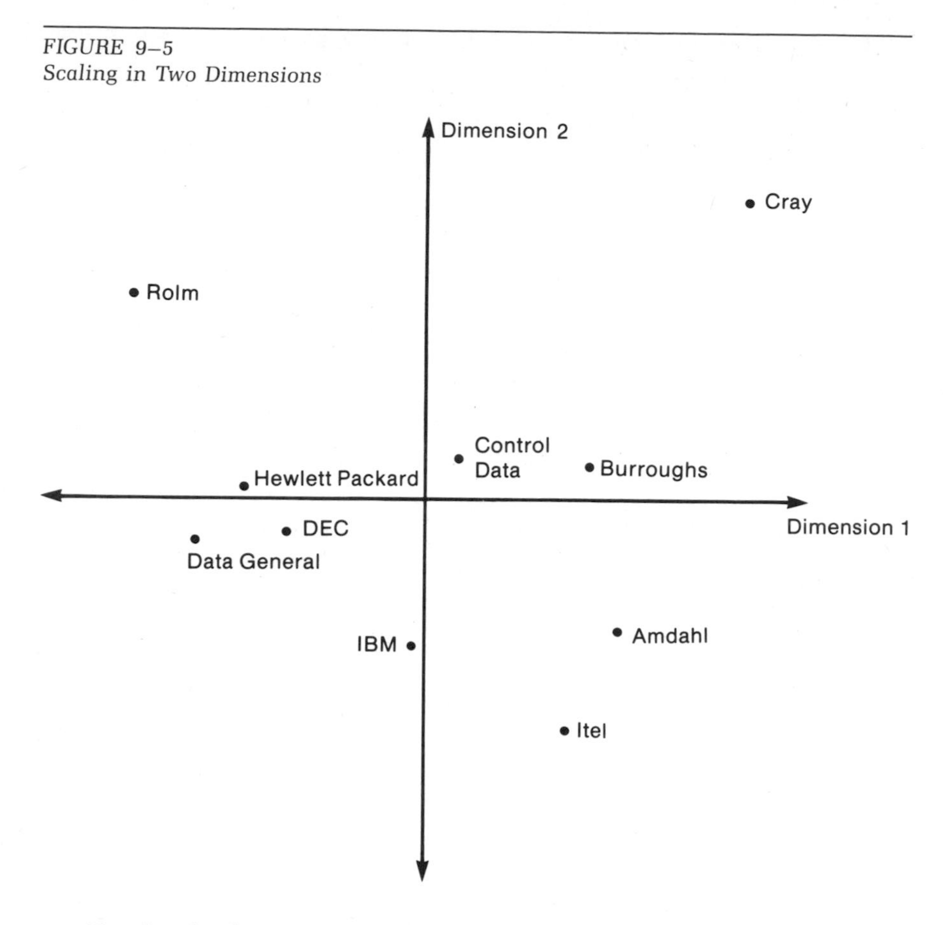

To clarify these points, we now turn to an interpretation of the output shown in Figures 9–4 and 9–5. In order to define measures of fit, we then turn to a more detailed consideration of how program KYST performs nonmetric scaling.

INTERPRETATION OF FIGURES 9–4 AND 9–5

Often in studies involving multidimensional scaling the interpretation of the geometric plots is done by the analyst rather than by the subject whose judgments underlie those plots. This practice seems to be wasteful of information, for often a subject is best able to explain the axes of an output plot by considering what criteria s/he could have been using in making the input judgments. In some cases, to be sure, subjects will be uncomfortable in dealing with geometric representations, but at least for those subjects who can deal with graphs it seems appropriate to involve the subject in the interpretation procedure.

The subject whose judgments were used for Figures 9–4 and 9–5 was very comfortable in working with geometric representations and, accordingly, he was asked to interpret the output, with some help from the analyst. The subject was asked to begin with the three graphs in Figure 9–4. (He was not shown Figure 9–5 until he had finished interpreting Figure 9–4). After some thought, he identified dimension 1 as running from companies known for "big number-crunchers" (Cray) to those known for mini- and microcomputers (DEC, Hewlett Packard, Data General, and Rolm). He commented that he had not been absolutely consistent in those judgments and thought, after examining the output, that Burroughs was a bit too far to the left, but that representation along dimension 1 was quite good.

He then tried to consider dimensions 2 and 3, commenting that he was finding their interpretation quite a bit harder. He finally decided that he could interpret dimension 3 as representing the breadth of the markets of the firms. He remarked that the plot seemed to show some errors or inconsistencies on that dimension and that both DEC and IBM should be placed lower, toward the broad market end, while Cray should be a bit higher, toward the narrow market end. He did, however, feel that the dimension was a useful one and that it was in fact an attribute he had used in making the input judgments.

Turning next to dimension 2, the subject decided that the second dimension described the hardware architecture (type of design) that the firm used. At one extreme were IBM and two firms whose architecture was extremely close to the main IBM designs (Itel and Amdahl). At the other extreme were Cray, which makes large specialized number-crunchers, and Rolm, which makes ruggedized miniaturized computers for use in severe environments. He thought that the dimension might represent closeness to IBM architecture and that such a dimension was a useful one, given the dominant position of IBM in the computer field. Again, there were a few points whose placements bothered him. He thought that Control Data should be between DEC and IBM on axis 2, but overall he thought the positions along the second axis were good.

Finally, the subject commented that his plots showed his personal concerns and experience in the computer field. He remarked that others knowledgeable in the area could have considered dimensions which he had not—the extent of support offered by a company to help users customize software was a particularly good example.

Next, the subject was shown the graph in Figure 9–5. He quickly identified dimension 1 in that figure as basically the same as the first dimension in Figure 9–4 (big number-cruncher versus mini- or microcomputer) and the second dimension in Figure 9–5 as like the second dimension in Figure 9–4, describing architecture of the products. He made the same types of comments about the misplacement of a few points and the general acceptability of the positionings. Asked whether the third dimension in Figure 9–4 appreciably improved the description of the computer industry over the one in Figure 9–5, he said definitely yes.

This example shows how plots from multidimensional scaling can be used to explore a consumer's view of an industry. Were a computer firm to undertake such a study, analysts would want to consider the views of different buyers, influencers, and experts on the industry to try to understand how their firm and other firms were perceived.

MEASURES OF FIT

In the above example, the subject was able to interpret all three dimensions and expressed a preference for the three-dimensional representation. Even so, we would generally want to have some measure of how well the configurations fit his input dissimilarities. (We might also want to proceed to four or more dimensions.) In this section we develop a measure of lack of fit, called *stress*.

Suppose that we want to perform multidimensional scaling on a subject's input dissimilarities in a specified number m^* of dimensions. It is useful to consider a few details of the actual computer fitting procedure. That procedure is basically a search. The computer starts with some initial configuration of points (obtained in one of the ways described later). It then must determine how well the configuration fits the input judgments. If the fit is not extremely close, the program proceeds to move the points in an attempt to improve the fit. It then recalculates the stress and repeats the cycle.

Suppose that in the course of this procedure we have some configuration of points in the m^* dimensions. What this statement means is that we have m^* coordinates for each object under consideration. With these coordinates we can calculate the distances between all pairs of objects. Suppose what we call the distance between object i and object j is D_{ij}. Recall that we have as input the judgmental dissimilarities between all pairs of objects. Suppose what we call the input dissimilarity for object i and object j is S_{ij}. Our problem is to express quantitatively the fit between the D_{ij} and the S_{ij}.

Because the S_{ij}s were assumed to contain ordering information only, it might seem reasonable just to count (in some way) the number of discrepancies in the ordering of the dissimilarities and the ordering of the distances. Such a procedure is not entirely satisfactory. To see why not, consider the two sets of distance values in Table 9–3. In each case, there are the same number of mismatches with the inputs. In each case, the 4th and 5th distances are out of order, as are the 8th and 9th. Yet, in the first set of distances, only relatively small changes would be needed to make the distance ordering consistent with the input judg-

TABLE 9–3

Input Similarities*	First Set of Distances	Second Set of Distances
1	1.1	1.1
2	1.3	1.3
3	4.1	4.1
4	4.6	5.6
5	4.5	4.5
6	5.7	5.7
7	5.8	5.8
8	6.4	7.8
9	6.2	6.2
10	8.1	8.1

*1 = Most similar.

ments, while in the second set those distances that are off are off by larger amounts. Since we want to interpret distances in the output plots of the scaling program, we would do better to define a measure of lack of fit that considered how far off the distances were.

Therefore, the following measure is used. The algorithm finds values, which we will call δ_{ij}, which agree as closely as possible with the D_{ij}s, subject to the constraint that they agree with the S_{ij}s in rank order. In this definition, agreement between the D_{ij}s and the δ_{ij}s is measured in terms of the sum of squared differences.[7] Also, agreement between the δ_{ij}s and the S_{ij}s means that if a particular S_{ij} is less than another S_{kl}, then the corresponding δ_{ij} must be no larger than δ_{kl}. (δ_{ij} must be less than or equal to δ_{kl}.)[8] For now we will assume that there are no ties in the input dissimilarities. Treatment of ties will be considered later.

The suggested measure of stress is then the sum of the squared differences between D_{ij}s and corresponding δ_{ij}s divided by the sum of the squared D_{ij}s. In symbols, stress is

$$\frac{\Sigma (D_{ij} - \delta_{ij})^2}{\Sigma D_{ij}^2}$$

The stress will be positive, with smaller values indicating better fits. (Other formulas, which are slight variations on this one, are also sometimes used.) To understand this definition, the reader will need to know more about how the δ_{ij}s are found; the next section pursues the topic.

[7]The choice of squared errors is made for computational convenience rather than for any theoretical reason.

[8]It might seem preferable to require that δ_{ij} be strictly less than δ_{kl}. Doing so would require that the user specify a minimal acceptable difference between the two (or else the computer could make the difference arbitrarily small). Instead, it is considered preferable to use the less restrictive requirement in the text.

MONOTONE REGRESSION

The procedure by which the δ_{ij}s are found, for a given set of D_{ij}s, in the most usual method for scaling is called *monotone regression*. The procedure can be demonstrated by example. Suppose that we have the set of input dissimilarities and distances given in Table 9–4. We want to find δ_{ij}s that are close to the D_{ij}s but that agree with the S_{ij}s in rank order. For the first three points there is no problem—we can simply make δ_{ij} equal to D_{ij}, since the D_{ij}s are the same order as the S_{ij}s. At the 4th and 5th dissimilarities we encounter trouble, because the distances are in the opposite order to the dissimilarities. With a squared-error criterion, the best choice is to make the δ_{ij}s equal to the mean of the out-of-order distances (4.4 and 4.0) at the two points. For the 6th and

TABLE 9–4

*Similarity** S_{ij}	D_{ij}
1	1.1
2	1.9
3	2.8
4	4.4
5	4.0
6	5.5
7	5.9
8	9.1
9	8.7
10	8.0
11	9.5
12	11.6

*1 = Most similar.

TABLE 9–5

S_{ij}	D_{ij}	δ_{ij}
1	1.1	1.1
2	1.9	1.9
3	2.8	2.8
4	4.4	4.2
5	4.0	4.2
6	5.5	5.5
7	5.9	5.9
8	9.1	8.6
9	8.7	8.6
10	8.0	8.6
11	9.5	9.5
12	11.6	11.6

7th distances, we can again make the δ_{ij}s equal to the D_{ij}s. The 8th, 9th, and 10th distances are out of order. Again, the best choice is to set the corresponding δ_{ij}s equal to the mean of the distances. Finally, the last two δ_{ij}s can be equal to the corresponding distances. Table 9–5 lists the results. Figure 9–6 shows a plot of the S_{ij}s and the D_{ij}s. Figure 9–7 shows the same plot with the δ_{ij}s added.

FIGURE 9–6

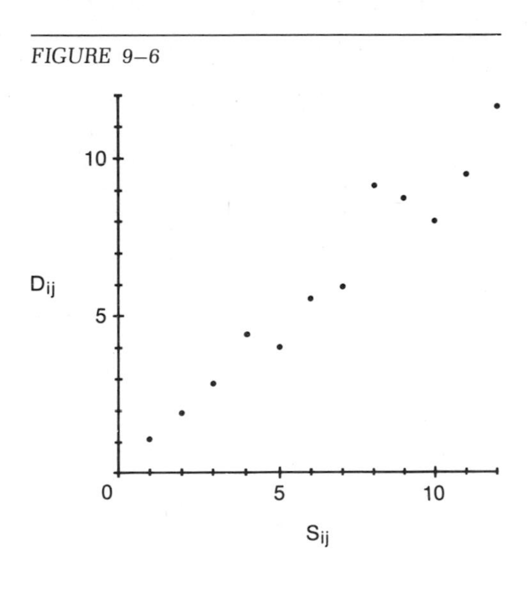

FIGURE 9–7

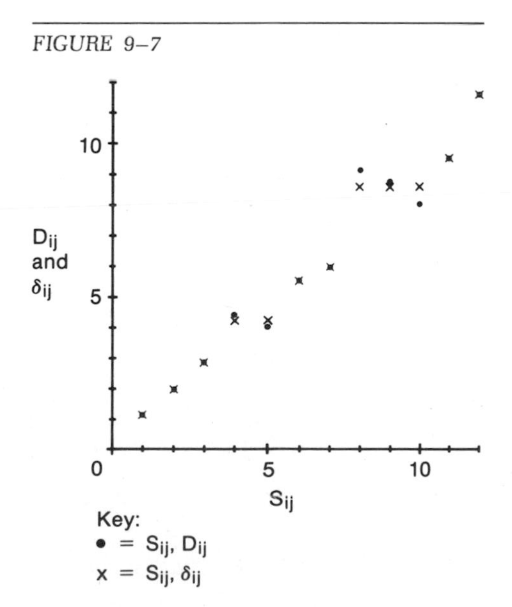

The monotone regression process by which the computer finds the δ_{ij}s is entirely analogous to the procedure that was just demonstrated. In understanding the results of scaling, it is important to note what conditions cause flat regions in the plot of the δ_{ij}s against the S_{ij}s. In the above fitting we were able to choose δ_{ij}s equal to the distances except where the distances were out of order. For out-of-order distances, we introduced flat regions in the graph of the δ_{ij}s. In general, flat regions have this meaning in multidimensional scaling plots of δ_{ij} versus S_{ij}. In some cases the situation can be extreme. For example, consider the dissimilarities, distances, and δ_{ij}s shown in Table 9–6 and plotted in Figure 9–8. Notice that the δ_{ij}s do in fact fit the D_{ij}s in the sense required in monotone regression. The stress is .0008.[9] Really, however, all that the fit has done is to divide the dissimilarities into one group of

TABLE 9–6

Similarities S_{ij}	*Distances* D_{ij}	δ_{ij}	$(D_{ij} - \delta_{ij})^2$
1	1.0	1.0	0.0
2	1.1	1.1	0.0
3	2.1	2.0	.01
4	1.8	2.0	.04
5	2.4	2.0	.16
6	1.7	2.0	.09
7	2.3	2.0	.09
8	2.2	2.0	.04
9	2.6	2.0	.36
10	1.6	2.0	.16
11	1.9	2.0	.01
12	1.4	2.0	.36
13	19.0	19.0	0.0
14	19.5	19.5	0.0
15	23.0	22.0	1.0
16	22.6	22.0	.36
17	21.8	22.0	.04
18	21.5	22.0	.25
19	22.5	22.0	.25
20	21.4	22.0	.36
21	22.2	22.0	.04
22	22.3	22.0	.09
23	21.7	22.0	.09
24	22.0	22.0	0.0
25	21.0	22.0	1.0

[9]The sum of the $(D_{ij} - \delta_{ij})^2$ values is 4.8 and the sum of the D_{ij}^2 values is 6,112. The ratio is .0008.

FIGURE 9–8

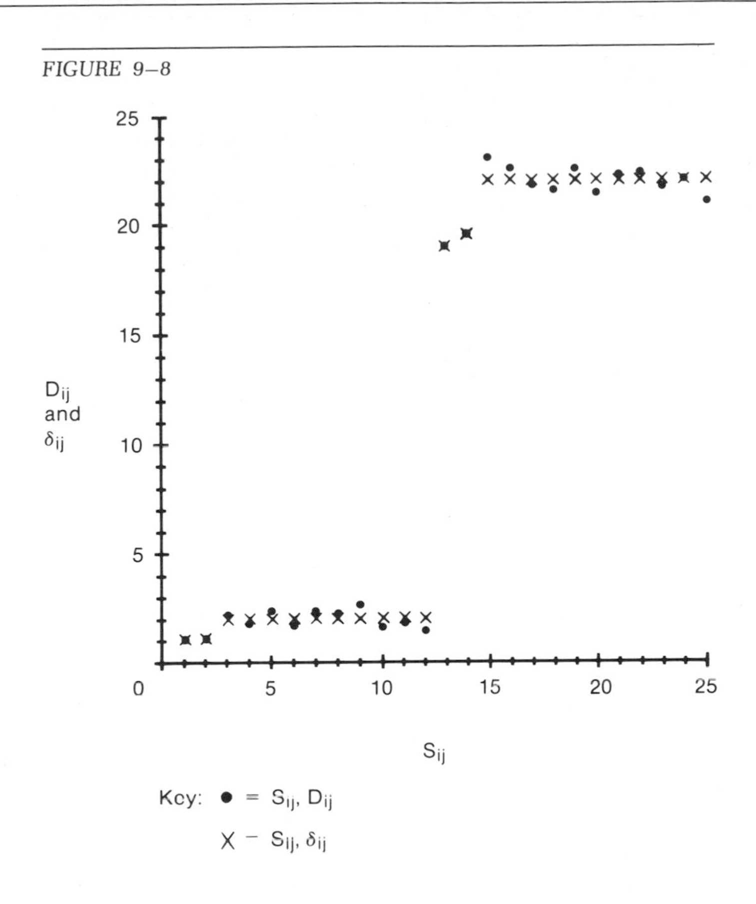

values that are low and another that are high. The distances and the dissimilarities do not match well at all.

Such situations are called *degeneracy*. In other words, long flat regions in the plot of the δ_{ij}s against the S_{ij}s indicate that the distances and the dissimilarities do not match well. In a way, the problem can be considered to arise because of the two-step fitting procedure. The algorithm involves a squared-error fit between the δ_{ij}s and the D_{ij}s and involves what is called a monotone fit (a fit in terms of order) between the δ_{ij}s and the S_{ij}s. The monotone fit is a rather loose restriction. δ_{ij}s that fit the S_{ij}s in the monotone sense can be close to distances that are not themselves close at all to the dissimilarities. In examining output from scaling problems, it is important for the analyst to consider both values of stress and also plots of δ_{ij} and S_{ij}. Figures 9–9 and 9–10 provide such outputs corresponding to the plots of computer companies in Figures 9–4 and 9–5. The fit in Figure 9–9, involving three dimensions, is closer than that in Figure 9–10. In each case, there are some flat regions in the graphs. Even so, the fits do provide some useful information with

FIGURE 9–9
Scaling in Three Dimensions (S_{ij}s, D_{ij}s, and δ_{ij}s)

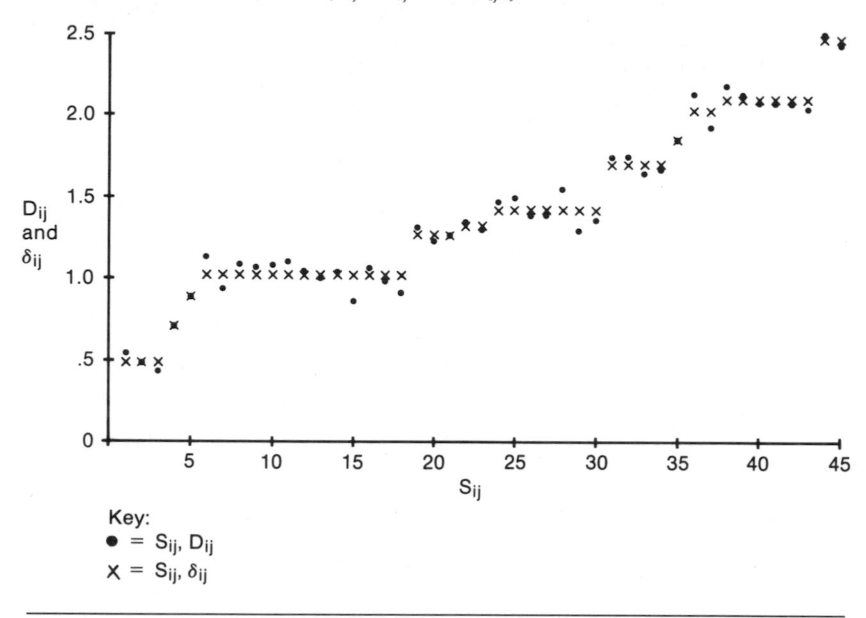

FIGURE 9–10
Scaling in Two Dimensions (S_{ij}s, D_{ij}s, and δ_{ij}s)

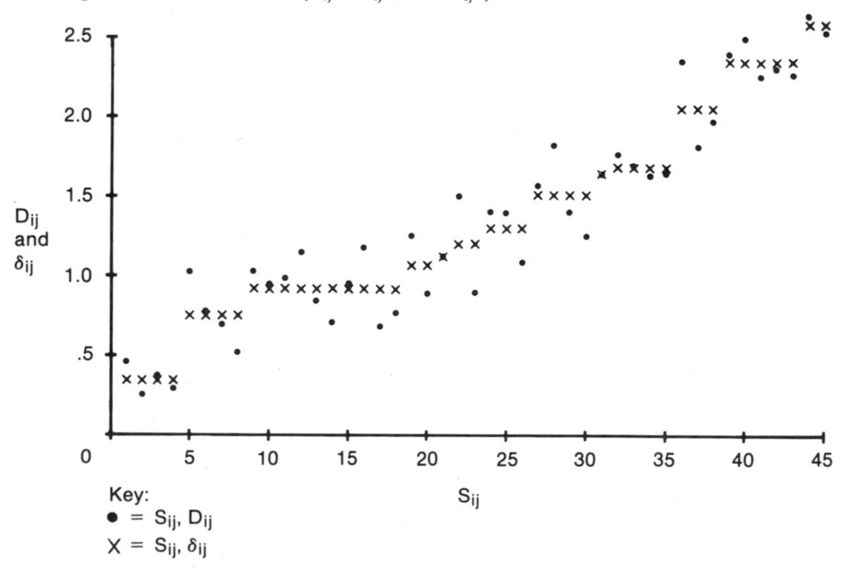

many distances increasing as dissimilarities increase. The stress values are .042 for Figure 9–9 and .105 for Figure 9–10. Tables 9–7 and 9–8 give listings of S_{ij}s, D_{ij}s, and δ_{ij}s.

We can now be more specific about how the multidimensional scaling procedures work. For any specified number of dimensions m^* they

TABLE 9–7
Three-Dimensional Analysis of Computer Company Data

Similarity S_{ij}	D_{ij}	δ_{ij}	Similarity S_{ij}	D_{ij}	δ_{ij}
1	0.53	0.48	24	1.47	1.42
2	0.48	0.48	25	1.49	1.42
3	0.43	0.48	26	1.39	1.42
4	0.71	0.71	27	1.39	1.42
5	0.88	0.88	28	1.55	1.42
6	1.13	1.02	29	1.29	1.42
7	0.94	1.02	30	1.35	1.42
8	1.09	1.02	31	1.74	1.70
9	1.06	1.02	32	1.74	1.70
10	1.08	1.02	33	1.64	1.70
11	1.10	1.02	34	1.67	1.70
12	1.04	1.02	35	1.85	1.85
13	1.01	1.02	36	2.13	2.03
14	1.03	1.02	37	1.93	2.03
15	.86	1.02	38	2.18	2.10
16	1.06	1.02	39	2.13	2.10
17	.98	1.02	40	2.09	2.10
18	.91	1.02	41	2.08	2.10
19	1.31	1.27	42	2.08	2.10
20	1.23	1.27	43	2.04	2.10
21	1.27	1.27	44	2.51	2.47
22	1.34	1.32	45	2.44	2.47
23	1.30	1.32			

select a starting configuration (in ways described in a following section). They calculate the stress. They then modify the configuration in what seems to be a promising way. The procedure involves what is called a gradient search. (The algorithm moves coordinates of the various points in directions that appear promising for reducing stress.) It then recalculates stress and proceeds with the search. The search can end under any of several conditions. If the stress becomes extremely small, the search stops. It also stops if successive steps do not seem to be resulting in further improvement. Finally, it stops if some prespecified number of steps have been taken.

Local Minima

As this description of stopping conditions may suggest, the nonmetric scaling algorithms stop either if they are making slow progress toward an optimal configuration or if the current configuration is better than all other configurations that are in some sense close to it. In technical terms, the algorithms stop if they have reached or almost reached local

TABLE 9–8
Two-Dimensional Analysis of Computer Company Data

Similarity S_{ij}	D_{ij}	δ_{ij}	Similarity S_{ij}	D_{ij}	δ_{ij}
1	0.46	0.34	24	1.41	1.29
2	0.25	0.34	25	1.39	1.29
3	0.37	0.34	26	1.08	1.29
4	0.29	0.34	27	1.57	1.51
5	1.02	0.75	28	1.82	1.51
6	0.77	0.75	29	1.40	1.51
7	0.69	0.75	30	1.25	1.51
8	0.52	0.75	31	1.63	1.63
9	1.02	0.92	32	1.76	1.68
10	0.94	0.92	33	1.69	1.68
11	0.98	0.92	34	1.63	1.68
12	1.14	0.92	35	1.64	1.68
13	0.84	0.92	36	2.35	2.05
14	0.71	0.92	37	1.81	2.05
15	0.94	0.92	38	1.97	2.05
16	1.18	0.92	39	2.39	2.34
17	0.68	0.92	40	2.49	2.34
18	0.76	0.92	41	2.26	2.34
19	1.25	1.07	42	2.30	2.34
20	0.88	1.07	43	2.27	2.34
21	1.12	1.12	44	2.63	2.58
22	1.49	1.19	45	2.53	2.58
23	0.89	1.19			

minima of the stress function. There is no guarantee that the local minima are in fact the best of all possible configurations (the global minimum). The only known way to address this problem is to perform the search from many different starting configurations in hopes that different starting configurations will lead to different minima, possibly only local minima, and that the global minimum will be included in these minima.

Starting Configurations

In part because of this problem of local minima the choice of starting configuration is important. One option in program KYST is for the user to specify a starting configuration by listing coordinates for all objects. In the rare event that the user has enough information to give a starting configuration, it is certainly appropriate that s/he do so. The second option is for the user to instruct the program to locate the points randomly in the specified number of dimensions. This solution is sometimes used repeatedly when the user is concerned about local minima. KYST is told to use a large number of different random starts.

The third choice for finding a starting configuration is for the program to use metric or quasi-nonmetric scaling to obtain a starting configuration. This choice is the standard one in KYST. Configurations found with this method are called Torsca configurations. The basic idea is that the input dissimilarities for nonmetric scaling are not exact distances but that they often contain considerable distance information and, moreover, that the metric algorithms tend to be quite robust. Therefore, those input dissimilarities are treated as if they are true distances[10] and subjected to metric scaling (or perhaps quasi-nonmetric scaling—which is metric scaling with "enhancement" steps) to obtain a starting configuration. Torsca configurations will in general be considerably closer to optimal configurations than will starting configurations found with the other options.

Number of Dimensions

It is usual in scaling to consider all numbers of dimensions between some user-specified maximum and some user-specified minimum. When the procedure is followed, the user must give KYST instructions for a starting configuration in the highest number of dimensions only. The program then proceeds to consider successively fewer dimensions until it reaches the minimum number. At any intermediate number of dimensions, the starting configuration is obtained from the ending configuration in the next higher number of dimensions. The procedure is simply to drop the last coordinate in the next higher dimensioned configuration. Because KYST usually is instructed to find principal axes,[11] with the first axis providing the most explanation of the dispersion of the points, the second axis providing the next most explanation, and so on, this procedure for dropping the last coordinate is a very sensible one.

In choosing among the results for different numbers of dimensions, the analyst must consider interpretability and fit as well. If more than two choices are considered, KYST will provide a plot of stress versus the number of dimensions. It is often suggested that bends or elbows in these plots suggest the correct number of dimensions. For example, in Figure 9–11 the graph drops sharply when the dimensionality is in-

[10]Thus, the ranks 1, 2, 3, . . . are treated as distances of 1, 2, 3, ...

[11]Scaling procedures must include some such convention for choosing axes because, for two possible axes, rotations of those same axes will do just as good a job of explaining dissimilarities. For further discussion of principal axes, see Chapter 6 on principal components analysis.

FIGURE 9–11

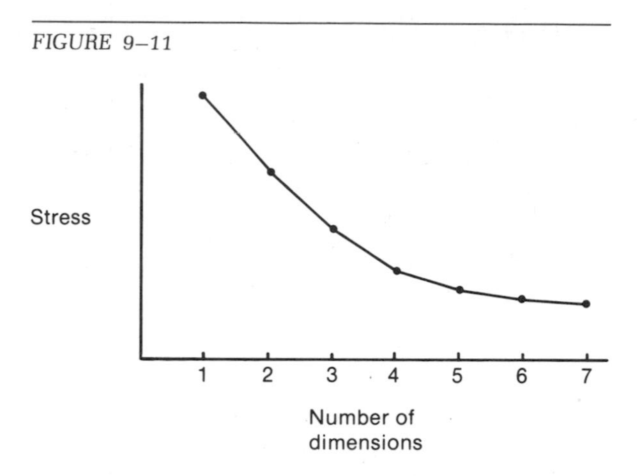

creased from 1 to 2, from 2 to 3, and from 3 to 4. After that, improvements in stress are considerably more modest. The graph suggests that 4 may be the correct number of dimensions. Before the final decision was made the analyst would also have to consider interpretability of results in the different numbers of dimensions.

FITTING PROCEDURES

The discussion of degeneracy emphasizes the weakness of the relation between δ_{ij}s and the S_{ij}s imposed by monotone regression. We would often prefer to impose a stronger relationship. KYST and other scaling programs do allow other choices (although the monotone regression one seems to be used most often). We would like in particular to insist that the δ_{ij} values change smoothly as the S_{ij}s change. KYST allows the user to require that the relationship between the two sets of values be a polynomial of any user-specified degree up to 4. Figure 9–12 shows hypothetical results for a polynomial of degree 2.

Notice that the relationship between δ_{ij} and S_{ij} is smooth. The fit between the δ_{ij} and the D_{ij} is still measured in a squared-error sense. Now, however, if the D_{ij} are close to the δ_{ij} they are close to a smooth function of the S_{ij}. Unfortunately, the choice of such *polynomial regression* cannot be guaranteed to give fits which are monotone—in other words, in which larger δ_{ij}s correspond to larger S_{ij}s. (Recall that our aim is to find a configuration with distances, the D_{ij}s, that correspond to the S_{ij}s. We want a greater input dissimilarity, or S_{ij} value, to correspond to a larger distance.) Figure 9–13 shows a hypothetical output graph in which the δ_{ij}s are a quadratic function of the S_{ij}s but in which both decreasing and increasing portions of the quadratic func-

FIGURE 9–12

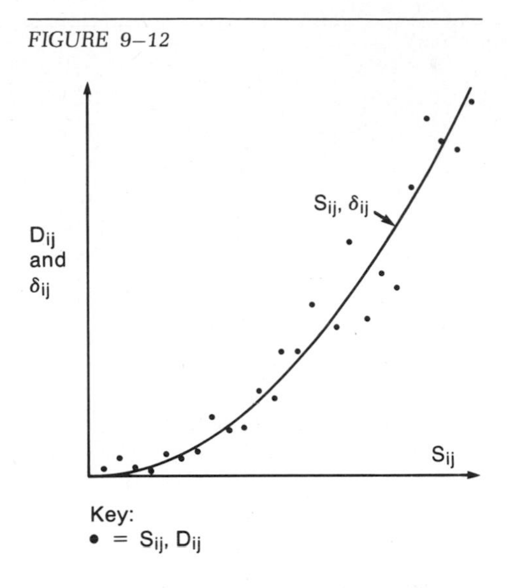

FIGURE 9–13

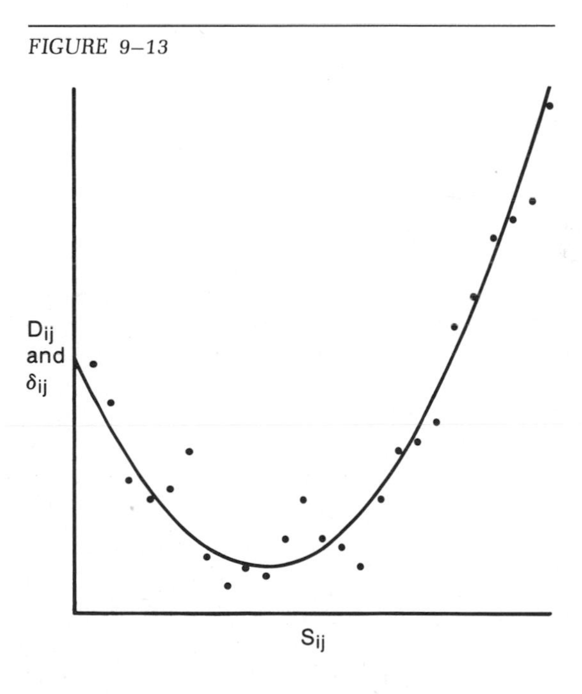

tion are involved. In the figure, the smallest values of D_{ij} and δ_{ij} correspond to intermediate, rather than small, S_{ij}s. One cannot in scaling require both monotonicity and smoothness. At best, the user can require one of these conditions and check for the other.

Treatment of Ties

In the preceding examples the subject never specified that two input dissimilarities were exactly equal. In other cases, such ties are allowed. KYST allows two different assumptions about the meaning of ties. The user may assume that the subject does not really mean strict equality but that s/he simply did not take the time or trouble to differentiate exactly between the two dissimilarities. Accordingly, the user can instruct KYST to assume that ties in the S_{ij}s do not imply anything about the δ_{ij}s. This choice is called the primary treatment of ties. Alternatively, the user can assume that the subject meant it when s/he said that two dissimilarities were equal and therefore can instruct KYST to assume that equality of two S_{ij}s should imply equality of the corresponding δ_{ij}s. This choice is called the secondary treatment of ties.

Number of Points Needed

The assessment task in preparing the input data for multidimensional scaling is difficult. (The reader should select 10 objects in some category, make up 45 cards[12] corresponding to the different pairs of objects and put the cards in order of increasing dissimilarity. S/he will quickly become convinced of the magnitude of the task.) As a result, there is often a temptation to use few objects. In nonmetric scaling, however, we are extracting distance or metric information from nonmetric input. Nonmetric inputs (orderings) are assumed to contain less information than would full metric input. To get meaningful metric output from nonmetric input in general requires considerable input information—in other words, a lot of points. There are no hard and fast rules. Documentation from Bell Labs, where many of the scaling procedures were developed, suggests that the bare minimum number of objects for 1 dimension is 5, that the minimum for 2 dimensions is 9 and that the minimum for 3 dimensions is 13.[13] These suggestions are based on empirical evidence rather than on theory. They suggest the difficulty of

[12]There are 45 possible pairs. (See pages 193–94.)

[13]The reader will note that the computer-companies data did not contain enough information for use of three dimensions under this strict rule (even though the resultant output seemed sensible). Four dimensions would clearly seem to be too many for those data.

obtaining enough input data for sensible fitting in large numbers of dimensions. (Because the primary treatment of ties assumes less information is contained in the input than is true for the secondary treatment, use of the primary treatment in situations with ties can exacerbate the problem.)

DISTANCE METRICS

All of the above analyses used the standard Euclidean metric in which the square of the distance between two points is the sum of the squared differences between the points on each of the coordinate dimensions. Other choices are possible. One is called the *city-block* or *Manhattan metric*. With this metric the distance is defined as the sum of the absolute values of the coordinate differences between the points. (The names for this metric come from the fact that this distance measure is appropriate for a city grid where one must move from point to point along the streets and cannot cut through blocks.) Another choice with some psychological justification is called the *dominance metric*. That metric defines the distance between two points as the maximum of the differences on individual coordinates; the argument is that subjects perceive the most noticeable (biggest) difference.

KYST can perform scaling with the Manhattan metric and it can work with something very close to the dominance metric. Unfortunately, however, investigators have found that non-Euclidean metrics are prone to give problems with degeneracy and with local minima. Those metrics raise serious possibilities of misinterpretation of output and hence their use is often dangerous.

INTERPRETATION OF OUTPUT

The computer-companies example showed a subject interpreting the dimensions obtained with program KYST. The earlier discussion mentioned that sometimes the investigator is more interested in groupings or clusterings of points than in the interpretation of the axes per se. The following example is taken from a review article on scaling written by R. N. Shepard,[14] one of the originators of the techniques. Figure 9–14 shows the plot obtained from judgments on dissimilarities of animal pairs. Shepard feels that this plot is most useful for identifying similar groups of animals, not for trying to interpret any axes.

[14]Roger N. Shepard, "Representation of Structure in Similarity Data: Problems and Prospects," *Psychometrika*, vol. 39:4 (December 1974) pp. 373–421.

FIGURE 9–14

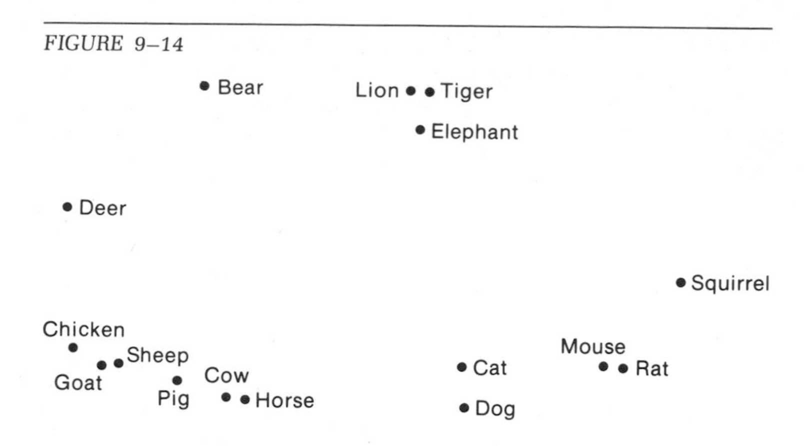

MULTIPLE SUBJECT STUDIES

Users of nonmetric scaling often want to consider the judgments of more than one subject. The above discussion has suggested what seems to be a sensible way of approaching the problem. Separate analyses can be performed for the different subjects. The subjects can help in the interpretation of the output corresponding to their judgments. The analysts can study the different interpretations and can then try to understand which types of subjects perceive the objects similarly. This description suggests the use of multidimensional scaling as an exploratory technique, one in which considerable amounts of investigator and subject judgment and interpretation are appropriate. As such, the technique can be very useful.

Unfortunately, it is more common for investigators to combine user judgments in a far more mechanical way. In fact, the most usual procedure seems to be simply to average the dissimilarities judgments of a group of subjects and then to input the averages to a scaling program. This technique seems at best ill advised. One must assume to start that the subjects perceive the objects according to the same set of underlying dimensions. In addition, the procedure requires that the subjects all make the same judgment about the extent to which a particular object possesses any particular attribute.[15] Finally, the procedure assumes that the subjects all assign the same importance to the different attributes in determining dissimilarities. These statements can be expanded a bit with an example. Suppose that we are willing to make the consid-

[15]Except for random error.

erable assumption that two subjects (X and Y) both perceive a group of five objects (A, B, C, D, and E) in terms of the same two dimensions, price and quality. Assume further that X and Y agree in their judgments of the degree to which each object possesses each attribute. For example, they agree that the price difference between A and B is the same as that between B and C. (As should be obvious, these assumptions are considerable.) Assume, however, that X is far more price-sensitive than

FIGURE 9–15

a. Subject X's Perceptions

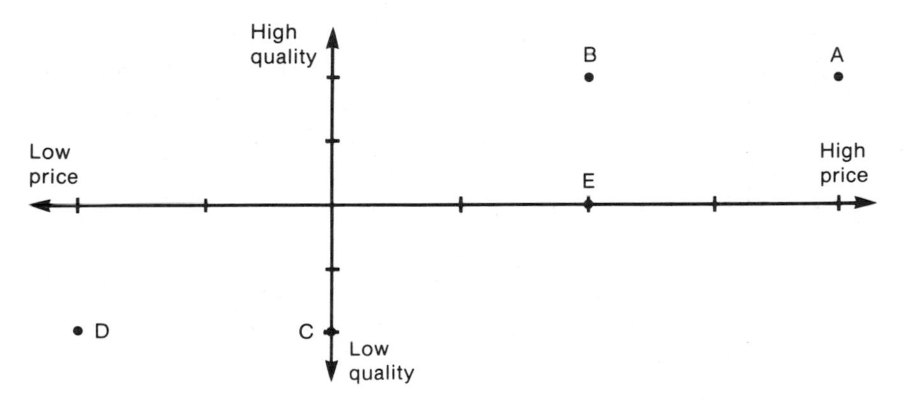

b. Subject Y's Perceptions

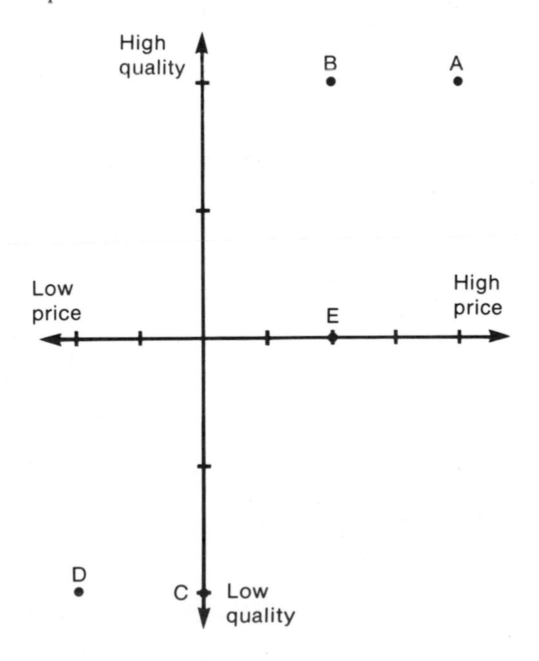

Y so that X's dissimilarities are much more strongly influenced by price differences than are Y's. Y is far more influenced by quality differences. Figure 9–15 shows two graphs that might correspond to X's and Y's perceptions. Notice that the two subjects disagree over the similarity of objects, even though they use the same basic dimensions and agree about the characteristics of the objects. For example, X finds B and E more similar than B and A, while Y finds the reverse is true.

In more general cases, we are not willing to make any of the above assumptions about shared perceptions. If so, the sensible solution is to analyze subjects separately. After preparing and interpreting maps for each individual, it may be sensible to try to group subjects according to similarities between their maps.

If in some situation we are willing to assume common underlying attributes and common perceptions about the extent to which objects possess attributes, then there is a possible approach that avoids assuming that subjects assign equal importance to the different dimensions. A program called INDSCAL, developed by J. D. Carroll and J. J. Chang of Bell Labs, can analyze such problems. The program takes input dissimilarity data from multiple subjects. It outputs information about the extent to which each object possesses each attribute and also information about the weights assigned by each subject to the different dimensions. In other words, it gives measures of the importance assigned by different subjects to the different attributes.

PREFERENCE MAPPING

Closely related to the nonmetric scaling procedures described above are procedures for what is called *preference mapping* or *unfolding*. Such procedures use input data in which subjects rank a group of objects in order of preference. There is one such ranking for each of a number of subjects. No subject compares objects directly. Each simply lists the objects from most to least preferred (or vice versa).

The unfolding procedures are based on the assumption that the subjects use the same underlying attributes in their perceptions and, moreover, that they agree as to the amount of each attribute possessed by each object. Finally, the subjects weight the attribute dimensions the same in judging distances. The subjects do differ in one respect, however. Each is assumed to have a single most-preferred point in the common configuration. In other words, each individual has a unique combination of attributes corresponding to his or her ideal object. Real objects are more or less preferred by the subject according to whether they are located closer to or farther from that subject's ideal point. The algorithms for preference scaling produce maps showing the locations

FIGURE 9–16

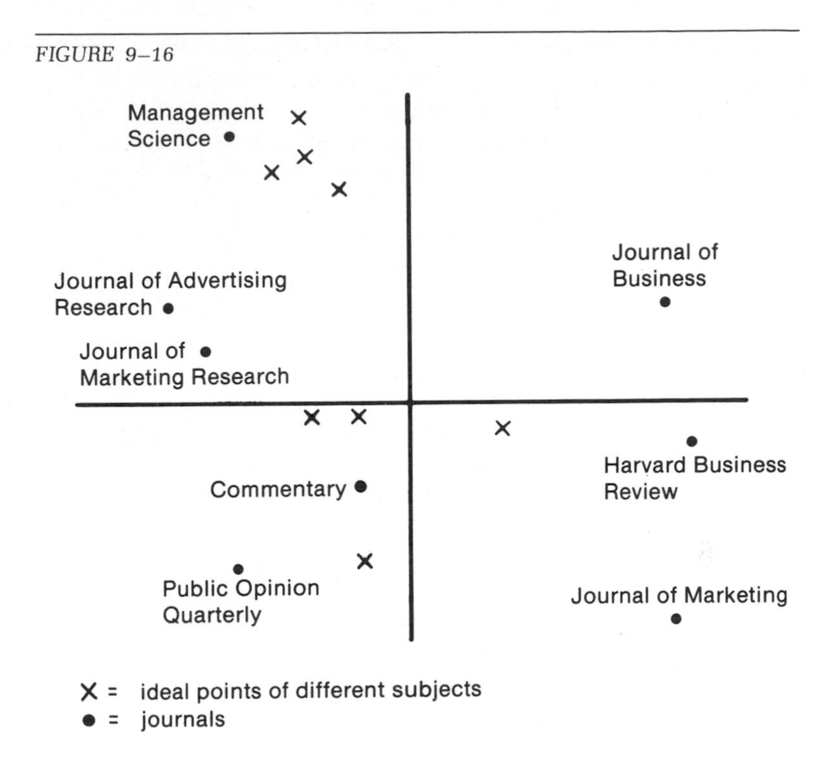

of the objects and the locations of ideal points for each subject. Figure 9–16 gives an example.[16]

The dangers in using such procedures should be obvious. The assumptions required are considerable, to say the least. The problems with assuming common perceptions have been mentioned above. The assumption that an individual has a single ideal point is also questionable. The subject who assessed the computer company data would prefer to deal with one company in some situations and other companies in other circumstances. As another example, consider the scaling of preference data on some food category, such as desserts. Studies using unfolding assume that a given individual has a single ideal dessert. In fact, however, people may like both fruit and chocolate cake, depending on how they feel. In fact, an individual may like chocolate cake and fresh strawberries yet dislike watermelon intensely even though s/he feels that in many ways watermelon and strawberries are similar.

[16] P. Green, F. Carmone, and P. Robinson, "Nonmetric Scaling Methods: An Exposition and Overview," *Warton Quarterly*, Winter-Spring 1968, pp. 28–41.

Finally, there are technical warnings about preference mapping. The preceding discussion explained how many data points are considered necessary in standard nonmetric scaling. The basic reason is that dissimilarities judgments contain less information than do distances; to convert nonmetric inputs into metric outputs requires considerable input data. The problem is much worse in preference mapping. For such procedures, the input data contain no direct comparisons of different objects by a single subject. Moreover, they contain no direct comparisons of the judgments of different subjects. The amount of input information is very low. In other words, the input information places very little constraint on the possible output configurations. It should not be surprising that the resulting configurations show very low values of stress. Despite such low stress, users should be wary of the results. The assumptions are extremely strong and the constraints imposed by the data are weak.

APPENDIX **A**

Bayes' Theorem and Likelihood

I. BAYES' THEOREM

Bayes' Theorem is a rule for manipulating probabilities (or frequencies). In particular, it is useful for incorporating new (or *sample*) information with previous (or *prior*) probabilities to form new (or *posterior*) probabilities that combine both sets of information.

As an example, suppose that we are considering a new test for components of some sort. The test indicates that the component is good or bad—but the test is not fully reliable. On the other hand, the new test is considerably less expensive than were previous methods. As a first step in evaluating the test we would like to find some conditional probabilities— which we denote $pr(CG|TG)$, $pr(CG|TB)$, $pr(CB|TG)$, $pr(CB|TB)$ (the probability the component is good given that the test says good, the probability that the component is good given that the test says bad, the probability that the component is bad given that the test says good, and the probability that the component is bad given that the test says bad, respectively).

Suppose we know from past testing that .8 of the components have actually been good and that .2 have been bad. The prior probabilities are thus

$$pr(CG) = .8$$
$$pr(CB) = .2$$

To calibrate the new test we might subject to it a group of components some of which we know (from the more expensive but more reliable test) are good and some of which we know (also from the reliable test) are bad. Suppose that we retest 50 good components and 50 bad components in this way and that we find that 35 of the bad components and 45 of the good ones test correctly. If we assume that this performance is representative of the new testing procedure, then we can write a set of probabilities for test results conditional on whether the components are good or bad:

$$pr(TG|CG) = 45/50 = .9$$
$$pr(TB|CG) = 5/50 = .1$$
$$pr(TG|CB) = 15/50 = .3$$
$$pr(TB|CB) = 35/50 = .7$$

Bayes' Theorem is simply a rule for converting this set of conditional probabilities (for test results given the true state of the component) into the other set of conditional probabilities (for true state given test results)—which are the ones we want (and will proceed to use, in ways not described here, to decide whether or not to use the new test).

In symbols, Bayes' Theorem for this problem says

$$pr(CG|TG) = \frac{pr(TG|CG) * pr(CG)}{pr(TG|CG) * pr(CG) + pr(TG|CB) * pr(CB)}$$
$$pr(CB|TG) = \frac{pr(TG|CB) * pr(CB)}{pr(TG|CG) * pr(CG) + pr(TG|CB) * pr(CB)}$$
$$pr(CG|TB) = \frac{pr(TB|CG) * pr(CG)}{pr(TB|CG) * pr(CG) + pr(TB|CB) * pr(CB)}$$
$$pr(CB|TB) = \frac{pr(TB|CB) * pr(CB)}{pr(TB|CG) * pr(CG) + pr(TB|CB) * pr(CB)}$$

Thus,

$$pr(CG|TG) = \frac{.9 * .8}{.9 * .8 + .3 * .2} = \frac{.72}{.78} = .92$$
$$pr(CB|TG) = \frac{.3 * .2}{.9 * .8 + .3 * .2} = \frac{.06}{.78} = .08$$
$$pr(CG|TB) = \frac{.1 * .8}{.1 * .8 + .7 * .2} = \frac{.08}{.22} = .36$$
$$pr(CB|TB) = \frac{.7 * .2}{.1 * .8 + .7 * .2} = \frac{.14}{.22} = .64$$

It is rather easy to become confused in applying the symbolic version of Bayes' Theorem and, therefore, often preferable to use an alternate (but equivalent) version of the same technique. Bayes' Theorem can also be presented in terms of *probability* (or frequency) *diagrams or trees*. Figure A–1 shows the probability tree for the initial known set of probabilities for this problem. The flow in the tree is from left to right. The branches at the right are reached via the ones on the left. Hence, the probabilities on the branches at the right are conditional ones—conditional on whatever event is represented by the branch at the left that was used to reach them.

FIGURE A–1

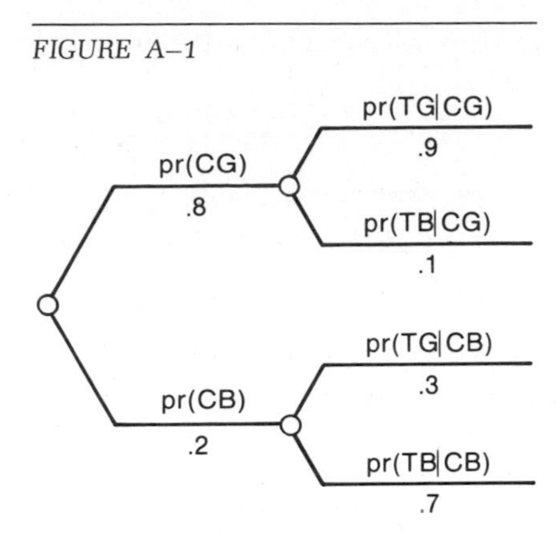

To apply Bayes' Theorem in such a tree, the first step is to label the endpoints with the products of the probabilities on the paths to those endpoints. Those values are *joint probabilities*. For example, the probability on the top endpoint in Figure A–1 would be the joint probability that the component is good and that the test says that it is good. Figure A–2 shows Figure A–1 with the endpoints filled in.

Next, we draw a tree with the order of the branches (and hence the order of the conditionality) reversed—in this case, we show test results to the left of the actual state of the component. We fill in all the probabilities that we can. First, we fill in the endpoints. Then, we fill in the marginal probabilities of the two test results by adding the probabilities at all the endpoints to which they correspond.

$$pr(TG) = .72 + .06 = .78$$
$$pr(TB) = .08 + .14 = .22$$

Figure A–3 shows the results so far.

FIGURE A–2

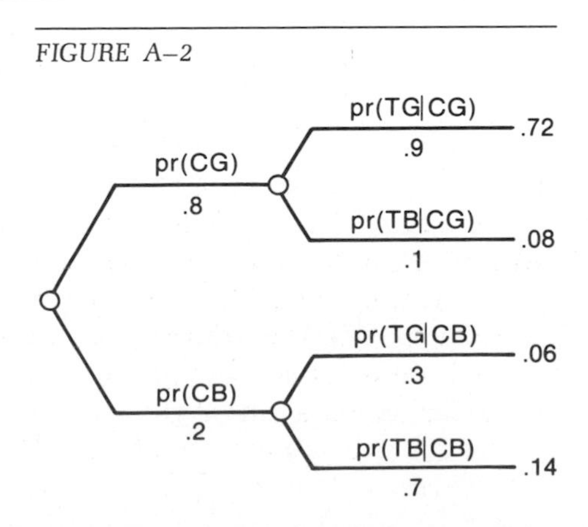

FIGURE A–3

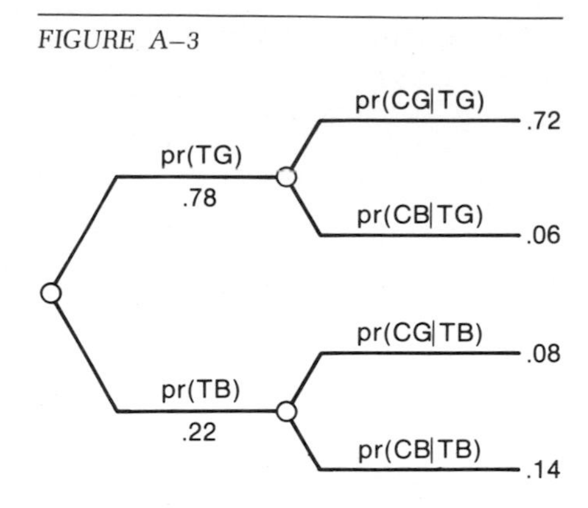

Finally, we find the desired set of conditional probabilities by noting that the probabilities along a path in the tree must multiply to give the endpoint value. Figure A–3 shows endpoints and one set of probabilities along the way. To find the desired conditional probabilities we simply divide (for example .72/.78 on the top branch and .14/.22 on the bottom branch). Figure A–4 shows the final result.

FIGURE A–4

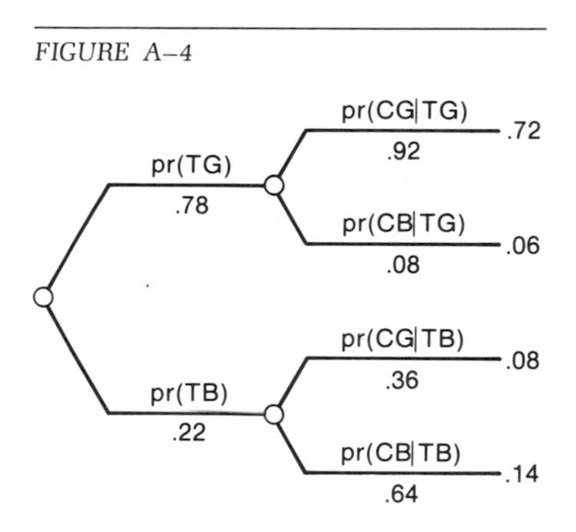

II. LIKELIHOOD

Likelihood is a concept from statistics that is important to understanding several techniques for multivariate analysis. This section describes the basic concept of likelihood, using an extremely simple example for illustration. Its purpose is to prepare the way for use of likelihood in the development of multivariate methods. This discussion assumes that the reader is already familiar with the mechanics of Bayes' Theorem.

Suppose that we are considering the introduction of a new product (a consumer packaged good) and that we are quite uncertain about the fraction of people in the target population who will try the product. For purposes of this simple example, suppose that we make the assumption that if we introduce the product either 40 percent of the target population will make trial purchases or else 10 percent will try the product. Clearly, in reality there are more than two possible levels of trial, but the mechanics of likelihood can be illustrated satisfactorily with only two. Suppose further that after careful consideration of the new product and of past introductions of analogous products we conclude that the higher trial rate is more likely. In fact, suppose that we believe at the outset that the probability of the higher fraction is .65. Then, the prior probability distribution for the fraction of triers is given in Table A–1.

The actual fraction of the target population who will try the product is unknown, however. Suppose we denote that fraction by p. It may be of substantial value to us to know more about the value of p without actually introducing the product. For one thing, we might choose not to introduce it if only 10 percent would try the product. Or, even if we would still introduce it if p took the lower value, we might make different choices for production facilities or for advertising and promotion, dependent on the values of p.

One way to learn more about the value of p without actually introducing the product is to question a sample of people from the target population and to determine whether or not each person in the sample would make a trial purchase. Suppose that we ask 10 such people whether they would try the product. Suppose further that we are willing to make the considerable assumption that if a respondent says s/he would buy the product then s/he would in fact buy it. In other words, we believe that the sampling method is able to elicit true purchase intentions. Finally, assume that of the 10 people in the sample, the second and the seventh say that they would buy the product while all of the others say that they would not. We might represent this result as a series of no's and yes's; NYNNNNYNNN. Given this sample information we would like to update our statement in Table A–1 about the probabilities of the two possible population fractions. Notice that a result of 2 yes's out of 10 is a bit more consistent with a true fraction of .10 then with a .40 fraction. On the other hand, a result of 2 in 10 is not unlikely with either possible fraction.

TABLE A–1

Population Fraction	*Probability*
.4	.65
.1	.35

In order to use this sample information we would like to calculate conditional probabilities: *Pr* (*p* is .4|sample result) and *Pr* (*p* is .1|sample result). In other words, we want to find the probability that the true fraction is .4, given the sample result, and the probability of a .1 fraction, given the same result. These probabilities are called *posterior probabilities* because they are posterior to the sample information. They should reflect both the information in the sample and the information in the prior distribution in Table A–1.

As a first step in using the sample information, we calculate what is called the *likelihood function*. The likelihood has a value for each possible population fraction. Its value for a particular value of *p* is defined as the conditional probability of the sample result, given the particular value of the population fraction. In symbols, the likelihood is

$$Pr\ (\text{sample result}|p).$$

Before proceeding to calculate values for the likelihood function, we should pause to consider the key issue of why such a function might be given the name likelihood. The basic idea is relatively straightforward but a bit hard to express. In essence, we argue that if the sample result is typical of the types of results we would expect for a particular value of the fraction *p*, then it is more likely that that value is in fact the true unknown value of *p*. On the other hand, if the sample result would be very unlikely for a possible value of *p*, then it is less likely that that particular possible value is the true value. Suppose, for the moment, that the sample had given five rather than two yes's. It is technically possible for such a result to come from a population in which only 10 percent of the people would try the product. Yet, a result of 5 in 10 is highly uncharacteristic of such a population. Therefore, if we had a sample of 5 yes's in 10 responses we would think it not very likely that the true fraction was .1. Similarly, suppose for the moment that we had a sample result of 0 yes's in 10 responses. It is technically possible but quite unlikely for such a result to come from a population of which 40 percent would try. Because such a sample result is so uncharacteristic of a population with true fraction .4, we would think it not very likely that the true fraction was .4. Note that this discussion has not said that population fractions for which the sample result is uncharacteristic are impossible; it has only said that such fractions are relatively unlikely. The measure of "characteristicness" that has been used implicitly is the conditional probability of the sample result, given a particular population fraction. Hence, the name likelihood is used for this probability function.

Returning now to the sample result in which the second and seventh people were the only yes's, we can proceed to calculate the likelihood function. For a population fraction of .4, assuming that the people were sampled entirely at random, the probability that the first person sampled would be a no is .6. The probability that the second person would be a yes is .4. Proceeding in this way to consider each of the 10 respondents, we can find the likelihood value of .0027:

$$.6 * .4 * .6 * .6 * .6 * .6 * .4 * .6 * .6 * .6$$

Similarly, consider the likelihood of a .1 population fraction. That value is .0043, as calculated below:

$$.9 * .1 * .9 * .9 * .9 * .9 * .1 * .9 * .9 * .9$$

In considering specific values of likelihood functions it is usually most useful to think about relative values. For the current values, the likelihood of a .1 fraction is roughly one and a half times as great as that of a .4 fraction. In other words, the sample result is somewhat more characteristic of the lower population fraction than of the higher one.

In the absence of any prior information, the likelihood might be converted into a posterior distribution. Such a procedure might be followed by classical statisticians who do not believe in using prior probabilities. It would also be the appropriate choice for the Bayesian investigator who thought at the start that the two possible fractions were equally likely. (In the terms of Bayesian statistics, such an investigator would be assuming a diffuse prior.) To convert the likelihoods to a posterior distribution, we would simply rescale them to sum to 1. The probability of the .4 fraction would be .27/(.27 + .43), or .39, and that of the .1 fraction would be .43/(.27 + .43), or .61. The resulting posterior distribution is given in Table A–2. We could also summarize the likelihood information by saying that the odds ratio was 27/43 for a .4 fraction over a .1 fraction.

In the current situation the probability distribution in Table A–2 would not be used. We had assumed the prior distribution given in Table A–1 and the information contained in that distribution must be incorporated into the posterior distribution, which should summarize the combination of the prior and the sample information. Notice that the prior distribution assigns a higher probability to the .4 value while the .1 value has a higher likelihood. The posterior distribution should be a compromise between the two sources of information.

Information from the two sources can be combined through Bayes' Theorem. In brief, the procedure is to find the joint probabilities (of

TABLE A–2
Posterior Distribution Based on a Diffuse Prior

Population Fraction	*Probability*
.4	.39
.1	.61

sample results and possible population fractions) and then to find the posterior probabilities *Pr* (.4|sample) and *Pr* (.1|sample). The procedure is demonstrated in Figures A–5 and A–6. The resultant distribution is given in Table A–3.

FIGURE A–5

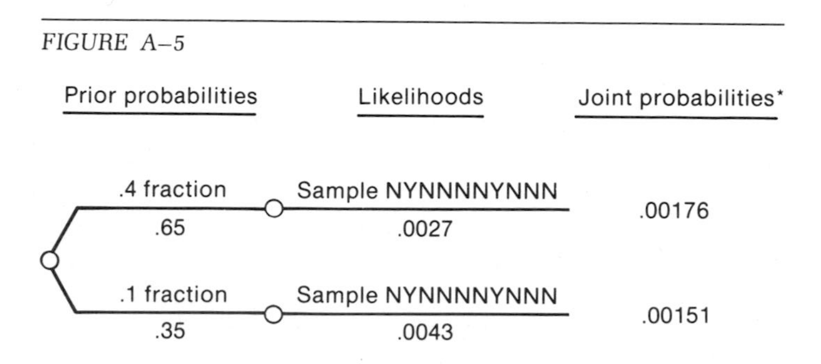

*The joint probabilities are found by multiplying prior and likelihood values.

FIGURE A–6

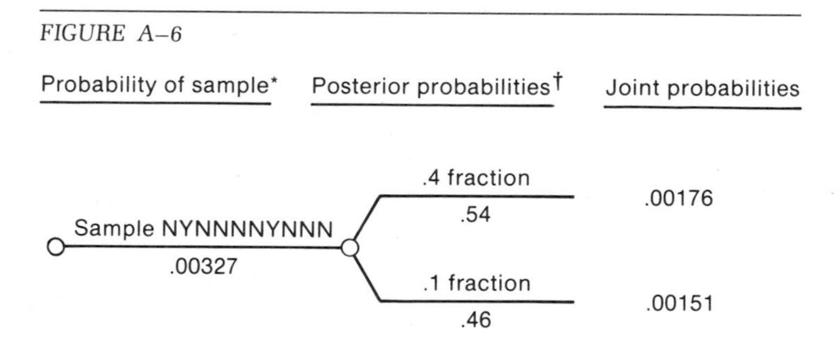

*The probability of the sample result is the sum of the two joint probabilities (.00176 and .00151) found in Figure A–5.

†The posterior probabilities are found by dividing the joint probabilities by the probability of the sample:

$$.54 = .00176/.00327$$
$$.46 = .00151/.00327$$

TABLE A–3
Posterior Distribution

Population Fraction	*Posterior Probability*
.4	.54
.1	.46

If we analyze an example a bit more realistic than the one discussed so far, we will almost certainly want to consider more than two possible values for the population fraction p. We might believe that all values between 0 and 1 were at least conceivable. If so, the likelihood function after a sample should be described for each value of p between 0 and 1. Figure A–7 gives an example. The height of the curve above a particular p value corresponds to the likelihood of that value.

If we had assessed a prior distribution for p, we would use Bayes' Theorem to combine the prior and the likelihood into a posterior distribution. On the other hand, if we assumed a diffuse prior (with no useful prior information), the posterior distribution and the likelihood function would have the same shape. For example, the value of p with the largest posterior probability would be that p corresponding to the highest point of the likelihood function. The value is called the *maximum likelihood* value of p.

The concept of likelihood is useful in situations in which sample information is obtained about measures other than population fractions. For example, it can be used in considering information about a population average (the average number of cups of coffee consumed per person in California is an example). The procedures will not be described here.

Likelihood and maximum likelihood are also useful in various multivariate methods, as mentioned at the start of this section. As an illustration, the following discussion sketches their use in regression.

Suppose that we have 30 observations in a database including a dependent variable y and two explanatory variables x_1 and x_2. The database is sketched in Figure A–8.

FIGURE A–7

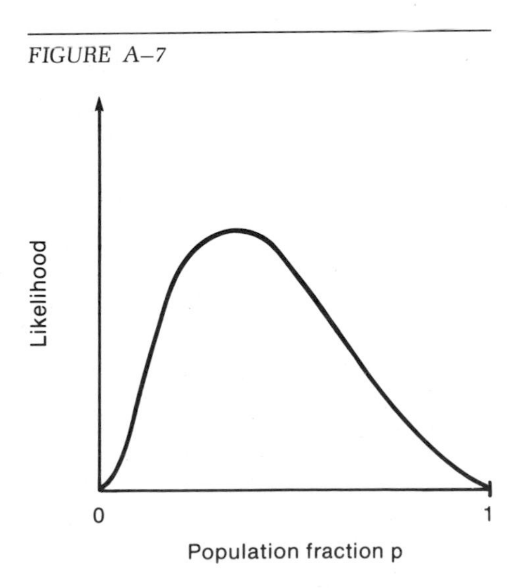

FIGURE A–8

Observation	y	x_1	x_2
1			
2			
⋮	⋮	⋮	⋮
30			

Suppose further that we want to fit a linear equation in x_1 and x_2:

$$y = \beta_0 + \beta_1 x_1 + \beta_2 x_2 + \varepsilon$$

Here β_0, β_1, and β_2 are the regression coefficients. ε is the residual, which takes a specific value on each observation. In fitting the model to the 30 observations, we want to find estimates of β_0, β_1, and β_2.

In the maximum likelihood approach to regression, we begin by making an assumption about the probability distribution from which the values of ε on individual observations are drawn. In general, we assume that the εs are drawn from the familiar bell-shaped normal distribution; the distribution is assumed to have mean 0 and standard deviation σ (which is not known).

The likelihood then enters in the choice of β_0, β_1, and β_2. For any possible set of estimates $\hat{\beta}_0$, $\hat{\beta}_1$, and $\hat{\beta}_2$ for the coefficients we can find the residual on each observation:

$$\text{Residual} = y - \hat{\beta}_0 - \hat{\beta}_1 x_1 - \hat{\beta}_2 x_2$$

Because the properties of normal distributions are well known, we can then find the probability density of obtaining those 30 residual values from a normal distribution:[1]

$$Pr\ (30 \text{ residuals} | \hat{\beta}_0, \hat{\beta}_1, \hat{\beta}_2)$$

This value is a likelihood measure. The 30 residuals are the sample result, analogous to NYNNNNYNNN above. The set of coefficients is analogous to the value of p above. We argue that a set of $\hat{\beta}_0$, $\hat{\beta}_1$, $\hat{\beta}_2$ for which the sample residuals are relatively characteristic is more likely to be the true set of coefficients than is an alternative set for which the residuals are uncharacteristic. If we assume a diffuse prior (no useful

[1]This probability is the product of the 30 individual values, one for each residual. Those individual values are obtained from a bell-shaped normal density curve.

prior information), then it is appropriate to select that set of regression coefficients which give the maximum likelihood value. The procedure for finding that set of coefficients is called *maximum likelihood estimation.*

Most readers will be familiar with least squares regression—the selection of values for the coefficients so as to minimize the sum of the squared residuals. Because of the special properties of linear functions and of the normal distribution, it turns out that the two approaches to regression give the same results. (The proof of this fact is considerably beyond the scope of this book.) Thus, the computationally easy least squares procedure can be used and will produce maximum likelihood results if the residuals are normally distributed. For other multivariate procedures there is no such easy short cut as least squares for obtaining maximum likelihood estimates. For such techniques, it is often necessary simply to search for maximum likelihood results.

In general, the basic ideas of maximum likelihood are the same for those other situations. We want to find a set of values, such as a set of regression coefficients. We make some assumptions about probabilities—such as the probability distribution of residuals. We then use the concept of likelihood and argue that a set of values producing a high likelihood is more apt to be the true set than is an alternate choice giving a lower likelihood value. Here likelihood of a set of values is defined as the conditional probability of the sample result, given the set of values. In the absence of other useful prior information, we select the set of values producing the maximum likelihood.[2]

[2]While using maximum likelihood values is a common practice in statistics, it is important to recognize potential pitfalls with that approach. One problem is a concern with all methods (including least squares regression) when they are used to find point estimates rather than distributions of possible values. In the case of maximum likelihood, the chosen point estimates may have a likelihood level not much higher than the levels for other sets of values. If those alternative sets give very different implications, including action implications, than does the maximum likelihood set, then accepting the single maximum likelihood set can be risky.

For some, though not all, of the applications of maximum likelihood estimation, there can also be potentially serious problems because the search procedure may find local rather than global (or overall) maximum likelihood values. Even in applications where this problem arises, it is serious for small databases but not for large ones.

APPENDIX B

Comparison of Techniques

The techniques for multivariate analysis described in this book can be contrasted with one another in several ways. Some of these distinctions are especially important to investigators in choosing appropriate methods in particular situations. This appendix presents some of the useful comparisons. Its first section draws some wide comparisons among the techniques in terms of types and scaling of variables, model assumptions, and so on. The information is expressed at length in the text and is summarized in Table B–1 at the end of this appendix. The second section draws more specific comparisons between pairs or small numbers of the techniques.

I. GENERAL COMPARISONS

A. TECHNIQUES WITH OR WITHOUT DEPENDENT VARIABLES

In many of the techniques, one variable is singled out as the dependent variable. The other (independent) variables serve to explain the dependent variable. In such cases, investigators want to explain (or predict) the behavior of the dependent variable by using the independent variables. Included in this group of methods are cross-tabs, additive cross-tabs, AID (automatic interaction detection), regression, binary regression, conjoint measurement, and discriminant analysis.

In other techniques the variables all serve the same role in the analysis; none is singled out to be explained by the others. Investigators want to explore joint relationships among a set of variables. Principal components analysis, factor analysis, and cluster analysis belong to this group.

Finally, multidimensional scaling involves only one input variable (which gives similarities—or dissimilarities—between pairs of objects). The technique is considered a multivariate one because it (usually) produces as output geometric representations with more than one dimension (or variable).

B. SCALING OF THE VARIABLES

In cross-tabs, AID, and additive cross-tabs the independent variables (perhaps after recoding) must each have only relatively few, distinct, levels, so that the observations can be sorted into cells on the basis of those variables. The independent variables can be dummy, categorical, ordinal, or cardinal. (For cardinal variables it is generally necessary to recode the variable by dividing its range of possible values into subranges, each defining a level of the new variable.) Most properly, the dependent variables should be cardinal (or, in some cases, dummy) variables, although the techniques are regularly used on ordinal variables as well.[1] The dependent variable may not be categorical.

Binary regression involves a binary (dummy) dependent variable and cardinal or dummy independent variables. (In practice, ordinal independent variables are sometimes also used.) In discriminant analysis with two groups the identifier (dependent) variable is binary (dummy). In discriminant analysis with more than two groups, the identifier variable is few-valued and typically unordered (categorical), with one value (or level) for each group being considered. In general, in discriminant analysis the independent variables are assumed to follow multivariate normal distributions within each of the groups under study and, thus, they should be cardinal. In practice, the technique proves to be robust and is used with independent variables that do not meet the normality assumption.

Standard regression involves a cardinal dependent variable and cardinal or dummy independent ones. In practice, ordinal variables are sometimes used (as dependent and/or independent variables).

In conjoint measurement, the dependent variable is ordinal, while the independent variables each take only a few different values representing the levels being considered for each attribute of the objects under study. Those few levels may be ordered (or even cardinal in nature), but they need not be. In nonmetric multidimensional scaling the single input variable is ordinal; in metric multidimensional scaling the input variable is assumed cardinal (with full distance properties).

---

[1]In this and other instances in which ordinal variables are used, although the techniques call for cardinal ones, investigators should be aware that the procedures will treat the variables as if they were cardinal. Users should be sure that they believe their variables are close enough to cardinal in behavior for their use to make sense.

Factor analysis and principal components analysis are generally used with cardinal variables. Ordinal variables, such as preference scales, are sometimes factor analyzed.

Finally, in cluster analysis investigators can use whatever types of input variables they want. It is also necessary, however, for users to specify a way to use values for the input variables to calculate distances. The resulting distance variable is assumed to be fully cardinal.

C. MODEL ASSUMPTIONS UNDERLYING THE TECHNIQUES

Most of the techniques involve models (such as the basic cross-tabs model that says the value of the dependent variable on an observation can be decomposed into a cell mean plus a residual). An exception is principal components analysis, which is simply a restatement of data (or, equivalently, a rotation of axes representing a set of variables).

The models underlying the procedures are discussed in the individual chapters on the techniques. This section highlights a few of the common model assumptions.

The techniques differ as to whether or not they assume additivity of the effects of the independent variables on the dependent variable. Cross-tabs and AID do not assume additivity. In the models for additive cross-tabs and for ordinary regression, the independent variables have additive effects on the dependent variables. In binary regression the effects are additive on the intermediate quantity (t) which is then transformed to give a value for the dependent variable. Similarly, in conjoint measurement there are additive effects of the independent variables on an intermediate value, which is related to the ordinal dependent variable through a transformation. Discriminant analysis assumes effects that are additive on the discriminant scores, or log odds of one group over another. The factor analysis model assumes that the general and specific factors should be added together (properly weighted) to give values of the manifest variables.

NOTE: In several of these techniques, transformed variables can be included to account for specific interaction (nonadditivity) effects. (Often, the new variables are defined as the products of two original variables). To be more precise, the preceding paragraph should state that techniques involve models that assume additivity with respect to the terms included in them—but that users can, as it were, "trick" the techniques into considering some specific interactions in the effects of an initial set of variables.

The techniques that assume additivity differ as to whether or not they assume linearity. Additive cross-tabs does not. Neither does conjoint measurement. In each case, the model involves a specific effect for each level of the independent variable, and there is no assumption that those effects change in a regular linear manner as the independent variable changes. In regression, binary regression, discriminant analysis, and factor analysis, the models involve linearity (as well as additivity) with respect to the independent variables. (Again, for some of the techniques, the linear model equations define intermediate quantities. And, to be more precise, we should note that the models involve linearity with respect to their input variables—but that those input variables may in fact be nonlinear transformations of other original variables.)

In two techniques—cluster analysis and multidimensional scaling—users must specify distance measures. The procedures then assume that dissimilarity is determined by the specified measure.

Finally, some of the procedures involve specific assumptions about probability distributions. In general, this book has focused as much as possible on data analysis rather than inference. In other words, it considered the analysis of specific sets of data rather than analysis followed by extensions of the results to larger groups of observations.[2] When investigators perform data analyses they need far fewer assumptions about probability distributions than are required for inference. There were, however, situations in the chapters in which distributional assumptions were made. In discriminant analysis, the independent variables are assumed to follow a multivariate normal distribution in each of the groups under study. In addition, in predicting group membership with discriminant analysis, users must give a prior distribution for group membership.[3]

II. SPECIFIC COMPARISONS

CROSS TABS, ADDITIVE CROSS-TABS, CONJOINT MEASUREMENT, REGRESSION, BINARY REGRESSION

Cross-tabs involves the simplest model, with the predicted value equal to a cell mean. No pattern of cell means is assumed; the model does not imply additivity or linearity. In large part because the model is so simple, cross-tabs generally requires many observations (enough to allow sound estimates of all the cell means).

---

[2]In a few cases, such as discriminant analysis, prediction for observations not in the original database is so integral a part of the technique that it was considered in the text.

[3]Even in pure data analysis, investigators must assume that their observations are sufficiently related for it to make sense to analyze them together. For example, investigators will generally remove aberrant outliers before analyzing a set of data. Some statistically oriented investigators may find it useful to think of this requirement in distributional terms.

The other techniques all impose more model assumptions than does cross-tabs. They do not require quite as many observations as cross-tabs. (In fact, users can think of the process as substituting assumptions for data.) Additive cross-tabs and conjoint measurement impose different forms of additivity. Regression assumes additivity and linearity (although variables can be nonlinear transformations of other variables).

Additive cross-tabs assumes additivity of the row and column effects. Conjoint measurement is much like additive cross-tabs but assumes only ordinal scaling of the dependent variable. This less restrictive assumption about the form of the dependent variable comes at the expense, however, of potential computational difficulties.

Additive cross-tabs gives different values (effects) for each of the levels of each variable. In contrast, regression produces a single coefficient for each independent variable. For each such variable, the coefficient is multiplied by specific values of the variable (and then added to other terms formed in a similar way) in calculating the predicted values.

Standard regression and binary regression differ basically as to the form of the dependent variable, which is cardinal in standard and dummy in binary regression. As a result of this basic difference, the techniques also differ in computational procedure and in the form and interpretation of their results.

DISCRIMINANT ANALYSIS AND BINARY REGRESSION

As explained in the chapter on binary regression, when the dependent (or identifier) variable takes only two values, in practice the two techniques often produce very similar results, despite the fact that their underlying models are very different. Discriminant analysis generally requires less computation than binary regression. As a result, discriminant analysis is frequently used as a substitute for, or else as a first step in, binary regression.

There is, however, one consideration that can dictate the use of one of these procedures rather than the other. The issue involves how observations were selected. In discriminant analysis users provide observations from each of the groups under study. They measure the values of the potential independent variables on each of those observations. In other words, users select values of the identifier variable and observe values of the independent variables. (The discriminant procedure assumes that the independent variables follow multivariate normal distributions within each group, while binary regression does not make that assumption.)

Sometimes, investigators want to select observations on the basis of values of the independent variables. For example, they might want to select observations so as to guarantee that wide ranges of values were included for the independent variables. Or, they might want to be sure to include observations with certain particularly interesting values on the independent variables. (At other times, of course, investigators select observations at random.)

When investigators select values for the potential independent variables, they should not use discriminant analysis. Instead, binary regression is appropriate; in binary regression, the independent variables are not assumed to follow a probability distribution.[4] On the other hand, when investigators have selected their observations systematically on the basis of values of the dependent (identifier) variable, discriminant analysis, rather than binary regression, is the appropriate choice.

GROUPINGS OF OBSERVATIONS: DISCRIMINANT ANALYSIS, CLUSTER ANALYSIS, AND MULTIDIMENSIONAL SCALING

Discriminant analysis and cluster analysis both deal with groups. In discriminant analysis, group memberships for the input observations are known and the procedure is used to find a rule for classifying (or, at least, for finding classification probabilities for) observations. In contrast, in cluster analysis the meaningful groupings are not known at the outset and the purpose of the technique is to help find such groupings (on the basis of a distance measure specified by the user).

Multidimensional scaling deals with configurations rather than groupings; it tries to find the relative locations of a set of objects in some specified number of dimensions. Users may want to subject the results of scaling to grouping (either through informal visual inspection or with formal cluster analysis), but the scaling procedures do not perform any grouping.

PRINCIPAL COMPONENTS ANALYSIS AND FACTOR ANALYSIS

Principal components analysis and factor analysis are sometimes confused, although they are really very dissimilar. Principal components analysis is a mechanical procedure for restating all of the information

[4]It is worth noting in passing that ordinary regression is like binary regression in this regard. Investigators may select values for the independent variables.

in a specific set of variables on a specific set of observations in terms of a new set of variables on the same observations; the technique does not involve a formal model.[5] In factor analysis, by contrast, users assume a model in which the values of a set of manifest variables are linear combinations of certain underlying but unobservable factors plus a component specific to each input variable.

The occasional confusion between the techniques likely arises for the following reasons. First, the equations used to express the two techniques look similar (but are not identical). Second, in the (very unlikely) event that in factor analysis the communalities are all 1, then factor analysis (with the same number of factors as variables) and principal components analysis give the same results. Third, one method of factor analysis (the principal factor method) does involve principal-component-type analysis (though only of that portion of the variables that is assumed to be explained by the common factors). Finally, or perhaps as a result of the previous reasons, investigators are sometimes sloppy in describing what technique they have used and what assumptions they made.

FACTOR ANALYSIS COMPARED WITH REGRESSION, CLUSTER ANALYSIS, AND MULTIDIMENSIONAL SCALING

Factor analysis attempts to explain a set of *observed* input variables by a linear combination of *unobservable* underlying common factors (plus specific components). In regression a single *observed* dependent variable is explained as well as possible by a linear combination of *observed* independent variables.

Factor analysis is used to examine the underlying factors, or internal structure, of a set of variables.[6] In contrast, in cluster analysis objects (variables or observations) are grouped in terms of overall similarity; there is no attempt to look inside or decompose the objects. There seems in practice to be considerable confusion between these techniques. For that reason, the appendix at the end of the chapter on cluster analysis compares the two techniques at more length.

Finally, factor analysis and multidimensional scaling are similar in that both are used to try to find underlying dimensions. Factor analysis finds dimensions underlying a set of input variables, while multidimensional scaling finds dimensions underlying a single measure (often a person's perception) of similarity or dissimilarity among objects. In

[5]It is true, however, that investigators may use principal components analysis in hopes that the first one or first few components will account for a large part of the variability in the full set of original variables.

[6]The technique is occasionally used on observations rather than variables.

factor analysis, the input variables are assumed to have full cardinal properties, while in nonmetric scaling the input variable is assumed to have ordinal properties only. In practice, the emphasis in factor analysis is usually placed on numerical outputs (especially the factor loadings), while in multidimensional scaling geometric outputs are usually emphasized. It is entirely possible (and perhaps advisable), however, to consider geometric representations from factor analysis and numerical coordinate values from multidimensional scaling.

TABLE B–1

	Dependent Variable?	*Scaling of Variables*	*Additivity?*	*Linearity?*
Cross-tabs	Yes	Dependent: Cardinal, dummy, (ordinal)[1] Independent: Few-valued[2]	No	No
Additive cross-tabs	Yes	Same as cross-tabs	Yes	No
AID	Yes	Same as cross-tabs	No	No
Conjoint measurement	Yes	Dependent: Ordinal Independent: Few-valued[2]	Yes	No
Regression	Yes	Dependent: Cardinal, (ordinal)[1] Independent: Cardinal, dummy, (ordinal)[1]	Yes	Yes
Binary regression	Yes	Dependent: Binary (dummy) Independent: Cardinal, dummy, (ordinal)[1]	Yes	Yes
Discriminant analysis	Yes	Dependent: Categorical[3] Independent: Cardinal[4]	Yes	Yes
Principal components analysis	No	Cardinal	Yes	Yes
Factor analysis	No	Cardinal, (ordinal)[1]	Yes	Yes
Cluster analysis	No	Any[5]	†	†
Multidimensional scaling	*	Nonmetric: Ordinal Metric: Cardinal	†	†

*One input variable only.

†The specific way in which input or coordinate values are combined to give distances is determined by the user's choice of distance measure.

[1]Sometimes used, though should be used with care.

[2]May be *based* on a cardinal, ordinal, categorical, or dummy variable but may take only a few different values.

[3]One value per group.

[4]Multivariate normal. However, ordinal (and even dummy) variables are sometimes used in practice.

[5]The user must define a cardinal distance measure based on the input variables.

Index